The Joy of Thinking:
The Beauty and Power of Classical Mathematical Ideas
Part II

Edward B. Burger, Ph.D.
Michael Starbird, Ph.D.

THE TEACHING COMPANY ®

PUBLISHED BY:

THE TEACHING COMPANY
4151 Lafayette Center Drive, Suite 100
Chantilly, Virginia 20151-1232
1-800-TEACH-12
Fax—703-378-3819
www.teach12.com

Copyright © The Teaching Company Limited Partnership, 2003

Printed in the United States of America

This book is in copyright. All rights reserved.

Without limiting the rights under copyright reserved above,
no part of this publication may be reproduced, stored in
or introduced into a retrieval system, or transmitted,
in any form, or by any means
(electronic, mechanical, photocopying, recording, or otherwise),
without the prior written permission of
The Teaching Company.

ISBN 1-56585-770-4

Edward B. Burger, Ph.D.

Professor of Mathematics and Chair,
Department of Mathematics and Statistics, Williams College

Edward Burger is Professor of Mathematics and Chair of the Department of Mathematics and Statistics at Williams College. He graduated Summa Cum Laude from Connecticut College in 1985 where he earned a B.A. with distinction in mathematics. He received his Ph.D. in mathematics from The University of Texas at Austin in 1990. In 1990, he joined the faculty at the Mathematics Department at Williams College. For the academic year 1990–1991, he was a Postdoctoral Fellow at the University of Waterloo in Canada. During his sabbatical years 1994–1995; 1998–1999; and 2002–2003 he was the Stanislaw M. Ulam Visiting Professor of Mathematics at the University of Colorado at Boulder.

In 1987, Professor Burger received the Le Fevere Teaching Award at The University of Texas at Austin. He received the 2000 Northeastern Section of the Mathematical Association of America Award for Distinguished College or University Teaching of Mathematics, and in 2001, he received the Mathematical Association of America Deborah and Franklin Tepper Haimo National Award for Distinguished College or University Teaching of Mathematics. In 2003, he received the Residence Life Academic Teaching Award at the University of Colorado at Boulder. Burger was named the 2001-2003 George Polya Lecturer by the Mathematical Association of America. He was also the 2001 Genevieve W. Gore Distinguished Resident at Westminster College and the 2001 Cecil and Ida Green Honors Professor at Texas Christian University.

Burger's research interests are in number theory, and he is the author of over 25 papers appearing in scholarly journals. He is the author of several books. Together with Michael Starbird, he co-authored *The Heart of Mathematics: An invitation to effective thinking* which won a 2001 Robert W. Hamilton Book Award. He also published *Exploring the Number Jungle: A journey into diophantine analysis,* and, with Robert Tubbs, co-authored *Making Transcendence Transparent: An intuitive approach to classical transcendental number theory*. He has also authored five virtual video CD-ROM textbooks with Thinkwell. Burger has served as Chair on various national program committees for the Mathematical Association of America; serves as Associate Editor of the *American Mathematical*

Monthly and a referee for many mathematical journals; and was a member of the Committee on Committees for the American Mathematical Society.

Professor Burger is a noted speaker and has given over 300 lectures around the world. His lectures range from keynote addresses at international mathematical conferences in Canada, France, Hungary, Japan, and the United States; to mathematical colloquia and seminars at colleges and universities; to presentations at primary and secondary schools; to entertaining performances for general audiences; to television and radio appearances including National Public Radio.

Michael Starbird, Ph.D.

University Distinguished Teaching Professor in Mathematics,
The University of Texas at Austin

Michael Starbird is a Professor of Mathematics and is a University Distinguished Teaching Professor at The University of Texas at Austin. He received his B.A. degree from Pomona College in 1970 and his Ph.D. degree in mathematics from the University of Wisconsin, Madison in 1974. In 1974, he joined the faculty of the Department of Mathematics of The University of Texas at Austin where he has stayed except for leaves as a Visiting Member of the Institute for Advanced Study in Princeton, New Jersey; a Visiting Associate Professor at the University of California, San Diego; and a Member of the Technical Staff at the Jet Propulsion Laboratory in Pasadena, California.

He served as Associate Dean in the College of Natural Sciences at The University of Texas at Austin from 1989 to 1997. He is a member of the Academy of Distinguished Teachers at UT. He has won many teaching awards including a Minnie Stevens Piper Professorship, which is awarded each year to ten professors from any subject at any college or university in the state of Texas; the inaugural awarding of the Dad's Association Centennial Teaching Fellowship; the Excellence Award from the Eyes of Texas; the President's Associates Teaching Excellence Award; the Jean Holloway Award for Teaching Excellence, which is the oldest teaching award at UT and is awarded to one professor each year; the Chad Oliver Plan II Teaching Award, which is student-selected and awarded each year to one professor in the Plan II liberal arts honors program; and the Friar Society Centennial Teaching Fellowship, which is awarded to one professor at UT annually and includes the largest monetary teaching prize given at UT. Also, in 1989 he was the Recreational Sports Super Racquets Champion.

His mathematical research is in the field of topology. He recently served as a Member-at-Large of the Council of the American Mathematical Society.

He is interested in bringing authentic understanding of significant ideas in mathematics to people who are not necessarily mathematically oriented. He has developed and taught an acclaimed class that presents higher-level mathematics to liberal arts students.

Recently, co-author Edward B. Burger and he wrote *The Heart of Mathematics: An invitation to effective thinking* which won a 2001 Robert W. Hamilton Book Award. A review in the *American Mathematical Monthly* (June–July, 2001) of this book said among much more, "This is very possibly the best 'mathematics for the non-mathematician' book that I have seen—and that includes popular (non-textbook) books that one would find in a general bookstore." He loves to see real people find the intrigue and fascination that mathematics can bring.

His previous Teaching Company course is entitled *Change and Motion: Calculus Made Clear*.

This course was made possible in cooperation with:

Key College Publishing, Emeryville, California
www.keycollege.com

Publishers of
The Heart of Mathematics: An invitation to effective thinking
by Edward B. Burger and Michael Starbird
http://www.heartofmath.com

Table of Contents

The Joy of Thinking:
The Beauty and Power of Classical Mathematical Ideas
Part II

Professor Biography i
Course Scope 1

The Amorphous Universe of Topology:

Lecture Thirteen	A Twisted Idea—The Möbius Band	4
Lecture Fourteen	A One-Sided, Sealed Surface—The Klein Bottle	20

From Paper Folding to the Infinite Beauty of Fractals:

Lecture Fifteen	Ordinary Origami—Creating Beautiful Patterns	36
Lecture Sixteen	Unfolding Paper to Reveal a Fiery Fractal	52
Lecture Seventeen	Fractals—Infinitely Complex Creations	68
Lecture Eighteen	Fractal Frauds of Nature	84

Measuring Uncertainty:

Lecture Nineteen	Chance Surprises—Measuring Uncertainty	99
Lecture Twenty	Door Number Two or Door Number Three?	115
Lecture Twenty-One	Great Expectations—Weighing the Uncertain Future	133
Lecture Twenty-Two	Random Thoughts—Randomness in Our World	152
Lecture Twenty-Three	How Surprising are Surprising Coincidences?	170

Conclusion:

Lecture Twenty-Four	Life Lessons Learned from Mathematical Thinking	187

Timeline 202
Glossary 205
Biographical Notes 210
Bibliography 215

The Joy of Thinking:
The Beauty and Power of
Classical Mathematical Ideas

Scope:

Fun, joy, pleasure, aesthetics, intrigue, beauty, richness, texture, power, and life are words not normally associated with mathematics. In this course, however, we will discover mathematics as an artistic and creative realm that contains some of the greatest ideas of human history—ideas that have shaped cultures. We will explore the fourth dimension, coincidences, fractals, the allure of number and geometry, and bring these weighty notions back down to earth.

Every lecture will develop incredible ideas by starting from commonplace observations and building from there. By counting the spirals on the prickly facades of pineapples and pinecones, we will discover a number pattern that has a life of its own and expresses itself within paintings, architecture, and music. By tracing the edge of a twisted strip of paper, we will develop insights into the shape of our universe. By having monkeys randomly typing on a keyboard to create *Hamlet*, we will discover the underpinnings of molecular motion. Thus in this course, sometimes frivolous, inauspicious beginnings quickly lead us to fundamental insights into our lives and our world.

Mathematical thinking leads not only to insights about our everyday lives and everyday world, but also points us to worlds far beyond our own. One of the joys of life comes from playing with ideas that are not bounded by mere reality. There is no reason why our minds cannot go where no mind has gone before and live to tell the exciting tale. A mathematical point of view can take us far and bring us back home with a new perspective on everything.

Whether viewers already enjoy mathematics or frankly hate mathematics, these excursions will enable everyone to bring great mathematical ideas to life. The lectures will create a lively and entertaining intellectual *tour-de-force* that opens eyes and opens minds.

One of the obstacles traditionally associated with mathematics is its hierarchical structure. One goal of these lectures is to demonstrate that mathematical ideas are within the reach of people who may not have an extensive background in mathematics. The universe of mathematics contains some of the greatest ideas of humankind—ideas comparable to the works of Shakespeare, Plato, and Michelangelo. These mathematical ideas helped shape history, and they can add texture, beauty, and wonder to our lives.

Our exploration of mathematical magic and majesty will begin most appropriately with the contemplation of the notion of number. Simple counting will lead us to surprising outcomes and the discovery of an incredible pattern of numbers in nature. Unexpectedly, exploring this natural number pattern will take us directly to a geometrical realm that influences the arts and our notion of beauty.

With our aesthetics as our guide, we will discover new mathematical insights through architecture, art, and even in snails. Issues of grace and elegance naturally give rise to an exploration of the idea of symmetry. Looking for the most symmetrical objects is an age-old quest with a rich reward. We will appreciate these wondrous objects not only for their individual character, but we will also see their intimate, dual relationships with one another. Even the mundane soccer ball will become a richer object when we see it in a mathematical light.

The fourth dimension, at first, appears inexplicable and inaccessible. However here we will apply the lessons of mathematical thinking to penetrate the haze and look into a universe with the eyes of the mind. Indeed, developing intuition about this invisible realm is a rich example of the power of analogy and abstraction. The abstract notions of the fourth dimension can even touch our hearts and minds through the beautiful works of art they inspired.

As we will see, seeking beauty and surprise need not take us outside our world. A simple creation from a strip of paper will offer us a glimpse into a world of bending and twisting space. Our explorations will take us far beyond our own world and intuition as we rethink the basic notions of inside and outside.

Other objects of beauty can exist completely in our minds, but can only partially be rendered owing to the limitations of reality. We will discover that fractal pictures are literally infinitely intricate. They arise from repeating a simple process infinitely often, and they seem to capture the complexity of nature. Repetition is at the heart of fractals and perhaps at the heart of our natural world. Surprisingly, these incredible vistas will then carry us far afield to the birth of computing machines. In fact, the simple act of folding a sheet of paper will offer us a bridge from the wondrous patterns of fractals to the very foundations of all programming.

A final theme of this course is that much of our world experience involves chance and randomness. Although counterintuitive at first, we will see that amazing coincidences are nearly certain to happen to all of us. After thinking about chance and randomness, surprises appear in a whole new light. Randomness is a powerful force—it permits us to develop a feel for measuring our expectations about future events whose outcomes are based on chance. We will be able to see lotteries, insurance, and other games of chance with a more critical and informed eye.

As we will see through the course, the mathematics offered here will not remind us of school—there will be no formulas; no problems; no equations; no techniques; no drills; and no, there won't be any tests. Some people might not even want to call it math, but we will experience a way of thinking that opens doors, opens minds, and leaves us smiling while pondering some of the greatest concepts ever conceived.

One of the great features about mathematics is that it has an endless frontier. The farther we travel, the more we see over the emerging horizon. The more we discover, the more we understand what we've already seen, and the more we see ahead. Deep ideas truly are within the reach of us all. How many more ideas are there for you to explore and enjoy? How long is your life?

Lecture Thirteen
A Twisted Idea—The Möbius Band

Michael Starbird, Ph.D.

Scope:

The surface of a Ping Pong ball, the surface of a doughnut, and the surface of a two-holed doughnut all have two sides: an inside and an outside. If we were bugs crawling on the outside surface and we returned to our starting point along any path whatsoever, then we would still be on the outside of the surface. Does every surface have two sides? Perhaps not.

To help us think about the two sides of this sidedness issue, we will build a physical model of a one-sided surface called the *Möbius band*. The Möbius band has been a crowd-pleaser for decades. After we have held a Möbius band and explored its endless edge, we feel an eerie sense of oneness: one edge, one side. We'll explore it through different experiments, including cutting it up in various ways. The Möbius band stretches our imagination and our intuition.

Outline

I. This lecture introduces a short section of the course dealing with "The Amorphous Universe of Topology."

   A. Topology is a branch of mathematics whose nickname is "rubber sheet geometry." The word topology comes from the Greek root *topos*, meaning "surface," and topology literally means the study of surfaces.

   B. In the last lecture, we learned about the fourth dimension, which is a concept of the mind created by exploring familiar ideas, such as a straight line, a plane, and three-dimensional space.

   C. In this lecture, we will again focus our attention on how mathematical strategies for thinking can be applied to our everyday lives.

   D. Specifically, we'll see how new ideas can be generated by generalizing familiar ideas.

   E. We will generalize the same familiar ideas in a different fashion. We think locally instead of globally.

II. Let's begin with a line, which as we know, goes on forever.
   A. Suppose you were a bug, and that line was your entire world.
      1. Locally, as a bug, you would see only a small part of the line.
      2. If the line were formed into a large circle, the bug would see no difference. It would still see only a small part of the line.
      3. This is another way to generalize the concept of a line: Instead of moving up in dimensions, we think of using the line to make objects that are locally similar to a straight line.
   B. Suppose, again, that you were a bug and that your entire world view was a plane.
      1. If you were living on a huge sphere, your view would still be planar.
      2. We are all familiar with this perspective because we live on a huge sphere, Earth. Locally, especially in Kansas, the Earth looks flat.
      3. Again, we are generalizing an idea. We have the same local impression, but the global structure is different.
   C. Can we think of any other surfaces that have this same feature?
      1. One example is a bagel. If you were living on the outside of a bagel, locally it would look essentially flat from any viewpoint. The same is true of an inner tube.
      2. In mathematics, a bagel or doughnut shape is called a *torus*. A double doughnut is called a *double torus*.
      3. All these surfaces look the same locally, but their global structures are different.
      4. Exploring objects that have this local feature is one way to form new ideas: What can be different globally when locally they are the same?
   D. Now, let's look at a cylinder. It is similar to the objects we looked at earlier, except that it has two edges, one on the bottom and one on the top.
      1. We can construct a cylinder by forming a circle with a long, rectangular strip of paper. We might also vary this construction slightly and explore the new features that we create.

2. If we make a half twist in the strip of paper before forming a cylinder, we construct a *Möbius band*.
E. We will explore the Möbius band by following the tenets we have learned thus far in the course.
   1. The first tenet we learned was to count.
   2. If we try to count the edges of the band, we find that it has only one edge.
   3. We also see that it has only one surface.
F. What happens if we cut the Möbius band in half down the middle all the way around?
   1. Before we make the cut, let's think about what the result will be.
   2. If we know how an object was constructed, we can sometimes analyze the constructed object more accurately if we look at the preconstructed version.
   3. We look again at a long, rectangular strip of paper and mark the points at which it will be joined to form a Möbius band.
   4. If we cut down the middle of the strip of paper, we would expect to get two strips, but as we see, in the Möbius band, the same points will still be joined together and we will have only one piece.
   5. When we actually perform the experiment, the result is a Möbius band that is twice as long and has two twists, but it is still only one piece.
   6. We have analyzed a finished product by looking at how it was constructed. We might use the same method to analyze other concepts, such as our society.
G. Let's perform another experiment with the Möbius band. This time, we cut the band very close to one edge.
   1. What will our results be when we complete the cut?
   2. If we cut the strip of paper before we make the Möbius band, we would have three pieces.
   3. However, in the Möbius band, what we are thinking of as the top strip and the bottom strip are connected; they make one strip that is twice as long as the original band.
   4. The center third, between the two strips, is not connected to anything. It will be a separate piece with a half twist in it.

5. By looking at the preconstructed object, we predict that after cutting, we will get two linked bands, one of the same length as the original with a single twist and one twice as long as the original with two twists.
   6. The band that has two twists is no longer a Möbius band, because it has two edges.
III. What is the difference between the Möbius band and a cylinder?
   A. If you were living on a Möbius band, locally it would look the same as a cylinder, sphere, or torus, but is there anything you could do on a Möbius band that you could not do on a cylinder?
   B. Suppose that you were a bug wearing a watch. We mark a clockwise arrow on the Möbius band to indicate the direction that your watch moves. Remember that you and your watch are completely flush with the band.
   C. If you move around the band and return to the point at which you started, your watch is moving in the opposite direction.
   D. The technical word for this concept is *non-orientable*. We can't find an idea of *clockwise* that is coherent for the whole object.
   E. The sphere, torus, double torus, and cylinder are all orientable.
IV. Möbius bands play some roles in the real world—in practical, symbolic, and artistic applications.
   A. Possibly, when Paul Bunyan cut down trees as a lumberjack, he used a Möbius band as a conveyor belt to transport the trees out of the forest. As he went deeper into the forest, he needed a longer belt; he cut his Möbius conveyer belt lengthwise down the middle and solved his problem.
   B. One of the important strategies we have learned in this lecture is to recycle our ideas, as we did with our understanding of the line and the plane. Indeed, the recycling symbol we see every day is a Möbius band.
   C. A final practical application of the Möbius band is to cut a Möbius bagel, which will be twice as long but won't be in two pieces, so you won't have to share it.

Suggested Reading:

Edward B. Burger and Michael Starbird, *The Heart of Mathematics: An invitation to effective thinking*, Key College Publishing, Section 5.2, "The Band That Wouldn't Stop Playing: Experimenting with the Möbius Band and Klein Bottle."

Questions to Consider:

1. Is it possible to have two Möbius bands of the same length situated parallel in space so that one hovers over the other exactly 1/8" away? Explain why or why not.
2. Take a strip of paper and put three half twists in it. Cut it lengthwise along the center core line. What do you get? Find an interesting object hidden in all that tangle.

Lecture Thirteen—Transcript
A Twisted Idea—The Möbius Band

Ed's last two lectures concluded the section of our course dealing with "The Visual World of Geometry." The next two lectures comprise a short section of the course entitled, "The Amorphous Universe of Topology." Now topology is a branch of mathematics whose nickname is "rubber sheet geometry" since when doing topology, we tend to bend and distort the surfaces we are studying, as you will soon see. Topology is actually my mathematical specialty, and when I was a student, my mentor explained to me the origin of the word "topology." He said that "-ology" meant "study of" and "top" meant "top," so he concluded that "topology" was the top thing to study. Fortunately, his topology was a lot better than his etymology. Actually, the word "topology" comes from the Greek root *topos* meaning "surface." So "topology" means literally "the study of surfaces." Although topology actually encompasses a broad study of abstract geometry in all dimensions, in these next two lectures, we will study surfaces just as the root of the word "topology" implies.

In the last two lectures, Ed took us on a journey to the fourth dimension. There we saw the power of using analogy as a means to create new ideas. In his lectures on the fourth dimension, Ed showed how we can take familiar objects such as a line, and a plane, and three-dimensional space and generalize them to create a new world—the fourth dimension.

I love the notion that we can devise specific strategies of thought that force us to think of new ideas. To me one of the most powerful lessons from mathematics is the lesson that we can somehow get ourselves to be smarter than we are, that is, we can get ourselves to think of things we have never thought before, just by following some effective techniques that help us generate new thoughts from the familiar. In fact, there are often many different possible generalizations of the same familiar ideas. In this lecture, we will again start with a line and a flat plane, but this time our new ideas will not involve going up in dimension. This time we'll generalize the line and the plane by thinking locally, and then seeing what we get globally. It's almost like a bumper sticker, "Think locally, wonder globally."

Okay, so let's think locally and start. Here we go. Suppose that you have a line. Well, one thing about the real line is that it goes on forever in both directions. Suppose you were a bug, though, who lived on that line, and when I say, "lived on that line," I don't actually mean on the line, I mean that the line was the totality of the bug's existence. All the bug can see is in this direction and that direction. It can't see out of the line. The line is its total world.

Well, then, if it lived on the straight line locally, it would just see a little, tiny part of the line. Suppose you took that line, though, and you folded it up into a circle. Well, if you imagine a very big circle, for example, the equator of the Earth, a very huge circle, and you thought about a bug that was living on that particular circle, locally, that bug would think that it was living on about the same place as if it were living on a line. In other words, just a little piece of that looks the same as a little piece of the straight line.

This, then, is another way to generalize the concept of the line. Instead of thinking of moving up in dimensions, we think of taking that line and making objects that are locally—that is, every point on it has a little neighborhood of it that looks like a straight line.

Well, for a straight line, a circle is about all you can get, so let's move on to one higher dimension, namely, dimension two. In the plane, then, suppose you were a bug, and your entire worldview was planar. All you saw was the plane. Okay? Now, let's view a sphere. This is a sphere, the boundary of a ball is a sphere. If you were a bug living on a sphere, then, locally, it would look very much like a flat plane, in particular, if it were a huge sphere.

Well, fortunately, we all have familiarity with a huge sphere, namely, the Earth, and we all have familiarity with flat spots, namely, Kansas. You see, then, if you are living in Kansas, the world looks very flat, because it's such a big sphere that locally it looks very flat. Okay?

That, then, is the concept of creating a world, a new idea, by taking an idea of the flat plane, and then saying, "Okay, could we construct new worlds in which we have the same local impression, but the global structure is different?"

Let's think about it. Can you think of some other surfaces, that are like the surface of a sphere, that have the same feature, so that, locally, if you were a bug just living on that surface, it would look exactly the same as if you were living on the boundary of the sphere or on a flat plane?

Well, every morning, if you happen to eat a bagel in the morning, you have some experience with such a thing because, you see, if you lived on just the surface of a bagel—not the inside of the bagel, for now, because you don't get to eat it; it's just about the surface—every point on the surface, even on the inside part, has a local neighborhood that looks like a sphere, and particularly if you think of really huge ones. Here's another one; like an innertube. You see? Even on the parts that are on the inside, curvy part, if this were a huge, huge bagel, and you were sitting right there, it would look basically flat. There's nothing distinctive about one point versus another. Okay? This is called the *torus*, by the way, and it's a fancy name; of course, this is math, and it's a fancy name. A torus is the surface of a donut.

Now, you could be even fancier, and look at this. Suppose this were smooth. This would be a double donut. Right? Or a *double torus*. This is an object that, if you lived on this, and this was just a smooth thing, rather than a ropey one, every point would look the same to you. Locally, then, we have surfaces that all look the same, but globally, the structure is different, and exploring the possibilities of objects that are similar locally, but different globally is a way to develop new ideas. You see what can be different, and when you have some feature that is the same, where do the differences lie?

We're going to put away the surfaces right here for a minute, and put them right here in this nice recycling bin, and instead turn our attention to surfaces that are like this surface, but in this case, we have a cylinder. This is the boundary of a tall can, and it's very much like the surfaces that we talked about before, except that it does have two edges to it, an edge on the bottom and an edge on the top. This allows us to expand our view a little bit, and think about surfaces where, if you were a bug on the inside, they would be the same, but you do have some edges to them, and this allows us to start thinking about a new way to construct a different idea, so here's what we're going to do.

The way you construct a cylinder, like that cylinder there, is to take a piece of paper—and I'm going to take one that's a bit longer, because it's easier to deal with, and this is not paper, but you take a long, rectangular object, and if you just brought it up and glued it together like this, you would get a cylinder. Right? Well, in mathematics, one of the things that you want to do, and I think in life, as well, is to do variations on a theme.

If you see an idea, some strategy that you can use to create something, in this case, a cylinder, if you alter it a little bit, you have a new world to explore. And we will find, by doing a slight twist on this construction, some really fascinating features, and the slight twist is to make a twist. That was a pun that was intended.

Here's what we do, then. We put a half-twist—make sure that you see what we're doing. You have a rectangle, and instead of bringing it around like this and gluing it, which would give us a cylinder, we're going to put a half-twist in it, and then we're going to glue it together. You see? By the way, there was double-edged tape on there; that's why it sticks together, for those who were curious about these practical matters.

First of all, isn't this object beautiful? It has a sensuous curve in it. It really is beautiful. This object is called the *Möbius band*, and it really is a wonderful construction.

Now, we're going to follow one of the most fundamental tenets that we have seen in this course of how to explore a new idea in more detail. The first tenet that we learned was to count. If you see an object, count things about it. Don't just be satisfied with a general impression, but actually count things.

We have something here. What is there about this that we could count? Well, the first thing we can do is to count the edges. Let's count the edges, okay? Let's just start here, and I'm going to put my fingers right here, and this hand I'm not going to move. This hand is right here on this edge. Now, watch what is going to happen. I'm going to move along this edge. You see it? So don't let me cheat. I'm just moving along the edge. I'm not letting go. I'm moving along the edge, moving along the edge. You see what's happening here? I'm moving along the edge. Oh, look. Do you see what's happening here? I'm on what you probably thought was the other edge, but you see, it is the same edge.

Now, watch. I'm going to continue to move. I am going to let go for just a second, but I'm going to come back and touch the same place, and continue along, continue along, continue along, continue along. See? No tricks, and I've got back to where I started, having covered the entire part. How many edges does this have? One. Well, now, that's sort of interesting.

We have something that has just one edge, but better than that, let's look at this. Suppose I take a marker here, and I start on the surface. Suppose that I'm an ant, and I'm crawling on the surface of this object. Here I am, an ant. See? I'll draw an ant. It's a beautiful ant, and suppose that my ant lover is on the other side. You see? Now, this is potentially a sad story, but look what happens. Fortunately, I'm going to draw the path that this ant can take, so here you go. You see how it's drawing? I hope you can see that, but if you can't see that, what you can do is to make one out of paper on your own and just do this. I'm just drawing, drawing without lifting the pen, I'm drawing. I do actually lift the pen, but I'm drawing. I just keep drawing, and after I've drawn this much, look what happens.

You see this? The ant was on this side, and without lifting my pen, the ant is now on the other side. In other words, then, I should not really say, "the other side," because the ant stayed on the side the whole time, so that really, there's only one side. This is an object that has just one side, and if we continued—you see, without lifting the pen, we've marked all the way around the whole thing; there we go, we're marking all the way around the whole thing. I'm just drawing a continuous line without lifting the pen. You see that everyplace I look has been marked. You see? It's blue over here, it's blue over here. Everyplace is marked, so that means that there's only one surface.

Now, this is quite interesting, because here we have an example of a surface unlike, for example, the cylinder. If you were an ant on the outside of the cylinder, and your true love was on the inside of the cylinder, you'd be stuck. You would have to go over the edge. That would be the only way you could get there, or if you were on a torus, or the boundary of a sphere on the outside, that's different from being a mole on the inside. Right? Okay.

Now, what I'm going to do now is to ask you to think ahead. I'm going to do an experiment with this Möbius band, and what I'm going to do is destroy it. Now, that's always fun. It's a destructive kind of era, so I'm going to destroy it, and the way I'm going to destroy it is to take a pair of scissors and just cut it in half, but I'm not going to cut it like this I'm going to cut it like this. Watch. I have to get started, and then I'm going to cut right down the middle, and then I'm going to cut all the way around. Here we go. Do you see this cutting? I'm cutting, cutting, cutting, all way around, and I'm going to keep cutting all the way around, but do you know what? I'm not going to finish the job right now, because I don't want to be destructive, and what I'm going to ask you to do is to think, to think about the question: What is going to happen? Can you predict what is going to happen when I finish that cut?

Now, the reason I want you to think about this is because it brings up a really valuable life lesson about the way to get insight into something, and that is that sometimes, if you know how something came to be, if you know the construction process by which something is made, you can sometimes analyze the effect of the constructed object better by looking at the previous reconstructed version. In other words, if we look at the blueprint by which this object is constructed, maybe we can think through why it is what's going to happen is going to happen. We can think it through.

Instead of continuing this cut, then—I will continue it later, and we will see what happens—I want to predict first, so let's look at an example of a pre-constructed one. Now, in this pre-constructed diagram, you notice that in order to tell us how we construct it, we have to say where the different points were glued. Okay? What I'm going to do is just to mark which points are glued to which, so that, for example, this point over here is glued to this point over here, because of the twist, so that tells me how to glue it. This is going to be twisted to make it correspond, Okay? You understand?

Instead of actually thinking about doing this as I did here in cutting, though, when I want to do is think ahead. Suppose I take this object. Now, I am going to be cutting it down the middle line, this centerline right here, all the way through, from the top to the bottom, and I'm trying to figure out what's going to happen, so let's think it through. We have this twist, we have this half-twist, and we're cutting down the middle. Suppose we cut down the middle before we construct it? Then, what would we have? We would have two strips, and how would they be joined to each other after we did the construction? What would be joined to what after you did the construction? Well, it's shown here. This point here is going to be glued to this point here. When we cut down the middle, how many pieces are we going to have? How many pieces are you going to have?

Now, think about it. Now, I really want you to think about it. Now, look. If this point is glued to this point, then they are connected to each other, right? This top part that you get after you cut down the middle then is connected to the bottom part, and then this point here is connected to this point. So when you cut down the middle, you're prediction is that unlike what you might have expected of getting two pieces, you should now expect to get only one piece.

Now, let's see if that's true. Now, in a way, the logic is so ironclad we don't actually have to do it, but just for fun, let's actually do it, just to see. Some people need the satisfaction of reality rather than just theory. That's the difference between an applied mathematician and a theoretical one. Right? We cut it, and look what we get. It has two twists to it, and is twice as long, and we know that is the case, because in the pre-constructed diagram, we actually had two strips, which, when assembled, make a strip that is twice as long, but a connected one. Okay? This, then, is advertising the concept of trying to analyze a finished product by looking at how that finished product was made, and we can look at all sorts of things in real life that way. You can look at how society was constructed, and see what forces caused things to come about the way they did, and use that as a tool for analyzing what the world is like, or what it could become like if the world had been certain ways.

Let's do another example. Were going to do another example, and this time, I'm going to ask you to think about it first, and then, see if you can correctly predict what happens. We'll take this strip again, and we're going to make a Möbius band. Once again, this is the pre-constructed diagram, and we make the half twist as indicated by the construction indications, and we glue it together to get another Möbius band. This time, what we're going to cut it again, but this time when we cut it, instead of cutting down the middle, we're going to cut it near one edge. You see? Here we are; we're nearer one edge than the other. We are never going to go to the middle, and we just cut along the whole way.

Now remember, there's only one edge, so after we have cut for awhile, pretty soon—we're going to hug this edge all the way around, and it's going to go all the way around the entire edge. For example, here, do you see what is happening? We are now near what we used to think was the opposite edge, but that we know is the same edge, but just halfway around the Möbius band, so that we continue around, and keep cutting. Now, we're going to ask ourselves the same question. What question are we going to ask? We are going to ask: What are we going to get? How many pieces are we going to get, and what will they be?

Well, once again, I'm not going to complete the cut, because we want to think first. Can we figure out what we're going to get after we complete the cut? See, I didn't quite complete the cut. Can we figure out what we're going to get by thinking about the pre-constructed diagram? Well, once again, these cuts are now one cut, but it's a long cut. It starts here, goes all the way across to here, and then it continues on, because, do you see this point here? This is the same as this point here, and so it continues along the bottom, so that that one long cut really goes twice across this.

Now, if we cut it before we constructed the Möbius band, we would have—it's easy, right? If you were to make those cuts, you would have three pieces, so now, let's just think about it. What are we going to get after we follow the construction rule of putting it in with a half twist, after we follow the construction rule, what will we get? Well, look at the top piece here, that top little thing. Is it connected to anything? Yes, because this piece up here, right here—do you see this corner right here?—is connected to this one down here, so that the top strip and the bottom strip are connected together. They make the long strip that is twice as long as the original band, just as when we cut down the center.

Look at the center piece, though, the center third. After we have cut the top to bottom, that center third is not attached to anything else, so that it will be a separate piece, and it will have the half twist in it. Can you envision that? What we actually predict, and before we actually do it, by looking at the pre-construction diagram, is that we're going to get two pieces: A long piece, and a thinner piece in the middle that's going to be another Möbius band. Let's see if it actually works.

Here we go. We complete the cut right here. Were almost done; we complete the cut, and look. It's sort of cool, isn't it? They are linked together. This one is just a Möbius band that is the same length as the first one. It's just the middle third of that first, original band, and then this one is twice as long, but has two twists. By the way, the one that's twice as long is not a Möbius band. It is not a Möbius band, because, for one thing, it has two edges. What used to be the edge of the Möbius band, the top, and the very bottom, is now one long edge, but the place where you cut is another long edge, so that this really is not a Möbius band, because if you put two twists in it, you don't get a Möbius band.

This, then, is an introduction to this wonderful object, the Möbius band. One question that we can ask about objects that have unusual properties is: What is the intrinsic difference between this Möbius band—that has this funny property, the way you cut it, and it has only one side to it and one edge—what is the intrinsic difference, if any, other than those that make it different from, for example, a cylinder? In other words, if you living on a Möbius band, and you were this bug again, your whole world was just the Möbius band, then locally, it with the same as on a torus, or on a sphere, or on a cylinder. Locally, then, it would look the same. Is there anything you could do actually on the Möbius band itself that would tell you that you were not on, for example, a cylinder?

That's the next question, then: Is there some intrinsic property of a Möbius band that makes it distinctive from these other objects? Let me illustrate something that is intrinsic to the Möbius band that makes it different.

Suppose you have a watch, okay? You are this bug, but you have a watch on, because time is very important. Here is a clockwise arrow that indicates which way your watch is going. Now, of course, this is the concept of something that is locally like a plane, and a plane, as you know, has no thickness at all, so that your whole world is—it's not like you are on one side of it, like this ant. This ant was actually crawling outside the band, but we're talking about literally being on the band, and you have this watch here, with a circle.

Here is an interesting feature: If you start moving around, so that your watch is going, it's telling time, and it's going, and you start walking around the band, so that here you come here, you know, your watch is still going, you see? You come to here, and your watch is still the same. You come to here, and your watch is same here. You come to here, and your watch is the same. No problem, right?

Now, I didn't draw these last ones, because I wanted to draw while we were doing it. Here we go; we're going along, and here is the next one, you see? I'm just trying the same, just what my clock would say, as I draw along, here we go. I'll draw one more, the same as that clock goes.

Then, look what happens. Do you see this one? The arrow is going the other way. You walk around your universe, and when you come back, you clock is going in the opposite direction. This object—I mean, maybe that's a way to turn back the hands of time, but we could hope that there's a Möbius band in our future—the technical word for this concept is that is *non-orientable*. You can't find a concept of clockwise that is coherent for the whole object, whereas for the sphere, the torus, the boundary of a doughnut, and the double torus, all of those objects that you are more familiar with, have the property that if you start drawing clockwise, and if you walk around, your clock is going to be the same when you get back to where you started. This, then, is an interesting concept called *non-orientability*.

Well, there are two other things I want to say about the Möbius band, and that is that a lot of people listen to these lectures, and they say, "Yes, but is it practical?" I don't like those people, incidentally, but some people say that, and then I figure, "Well, okay, I'll respond to these." Let me tell you a practical thing, then. This was something, and I think this is a true use of the Möbius band that was done by Paul Bunyan. Does that make it true? Here we go, then.

Paul Bunyan was, as you know, a lumberjack. Is that right? Good. Yes, he was a lumberjack, and so, when he would go into the woods and would start cutting down trees, of course, you've got to get the trees out of the woods. Well, now, when you start cutting down the trees, a good way to get trees out of the woods is to have a conveyor belt system. You can make a conveyor belt system, and here we go. We make a conveyor belt system like this.

However, Paul Bunyan is a very shrewd guy, and instead of having his conveyor belt as a regular conveyor belt, he made a conveyor belt that had a half-twist in it; that was the Möbius band. Here he goes, then; he is putting logs on, and they come out of the track, but of course, he has cut a lot of trees down, and so, pretty soon, he's deeper into the woods. Then, the problem is that he has to take the trees to the end of the conveyor belt, and how does he solve that problem? Well, you see, he has thought ahead, and he put that half-twist in it so that now, you see, what he can do is take his conveyor belt, cut it lengthwise down the middle, moves the ends of the rollers to twice the distance apart, and here you go. He's got a conveyor belt that goes deeper into the woods, and it's good for bringing the lumber much further into the woods, because it's twice as long. You see? True story; as true as Paul Bunyan is, anyway.

One of the important things that we've really learned here, though, is take some of our ideas and to recycle them, right? Because we've used them again. We used the idea of the plane, and recycled it. In fact, this is very important. If you look right here to the recycling symbol, if you look carefully at that, do you know what that is? A Möbius band. You have seen a Möbius band every day of your lives, then. The recycling symbol is the Möbius band. In fact, there is some beautiful artwork. M.C. Escher drew a picture of a Möbius band with ants crawling on it. That's why I referred to ants, and you'll see an artistic rendition of that that's very attractive, and many other works of art.

I want to finish with a very practical application of this idea, though, and that is, when you get up in the morning, and you have a bagel, and you want to cut this bagel, but you don't really want to share it, let me show you what you can do. You can start cutting it, but as you start cutting around, you rotate a bit, so that when you get halfway around, you are horizontal. You see? I started here, vertical, coming in from the top, and then as I cut around the side here, I rotated it until it was horizontal, and then I keep rotating, so that when I get back to the beginning, I'm now vertical again. Now, when this happens, what I have really made is a Möbius band, and look what happened. The bagel is completely connected, but twice as long, so that it's connected, and you don't have to share it anymore. This, then, is an example of a practical application of a Möbius band. Thank you.

Lecture Fourteen
A One-Sided, Sealed Surface—The Klein Bottle

Michael Starbird, Ph.D.

Scope:

The Möbius band was constructed by bringing the two ends of a strip of paper together with a half twist. The construction process gave us a one-sided surface, but unlike the sphere or torus, it has an edge. Could we construct a one-sided surface with no edge? Unfortunately, no one-sided surface without an edge can be constructed entirely in three-dimensional space. Nevertheless, we can effectively describe an elegant one-sided surface known as the *Klein bottle*. The rules for constructing a Klein bottle are simple, and the resulting surface can be attractively modeled and joyfully contemplated, even though it cannot actually be built. The neat property of a Klein bottle is that its inside is the same as its outside!

The Klein bottle has many interesting mathematical properties, but it is also an especially intriguing work of art. Stone, glass, and metal renditions of the Klein bottle have added grace and beauty to many museums.

Outline

I. As we saw in the last lecture, the idea of representing objects and exploring them turned out to be useful in creating new ideas.
   A. One of the most useful strategies in mathematics is to abstract ideas. We take a concrete idea and push it to different domains.
   B. In this lecture, we'll see that looking at representations and pushing them to the abstract will produce conceptual surprises and beautiful works of art.

II. Let's return to our "construction diagram" for a Möbius band.
   A. As you recall, to make a Möbius band, we join two opposite edges of a rectangular paper strip with a half twist in the middle.
   B. Can we apply a similar idea to some other familiar surfaces?
      1. For example, would it be possible to construct a sphere using the concept of joining the edges of a piece of paper together?

2. Let's start with a square piece of paper. We'll identify points on the edges of the paper that will be glued together to form a sphere.
3. In fact, if we were working with a thin sheet of rubber, we could glue these points together, inflate the resulting construction, and end up with a shape that resembled a sphere.
C. Let's try another shape—the torus.
1. A torus could be constructed by rolling up a piece of paper to form a cylinder, then joining the two ends of the cylinder together.
2. We must still specify which points on the edges of the paper will be joined together to form the torus. This is relatively easy, because there is no twist in this shape, as there was in the Möbius band.

III. In mathematics, we sometimes combine a familiar method, such as our method for constructing a torus, with another idea that we may have seen before, in this case, the idea of putting a twist into a constructed object, as we saw in the Möbius band.
A. In other words, can we combine the idea of gluing edges of a surface together to construct a new object with the idea of adding a twist?
B. If we identify the points that will be glued together to form a cylinder with a twist in it, we find that the points don't match up when we try to construct the object.
C. We could try various ways to get the points to match up, such as rotating the paper, but we can't quite make the construction diagram work. We need to think abstractly, even though we may have difficulty constructing the object physically.
D. If one end of our construction could pass through the surface of the other end, the points would match up perfectly.
E. The result would be an abstract object of the mind, the *Klein bottle*. We can look at a model of the Klein bottle, although it is not completely accurate.

IV. In a sphere or other familiar object, we can observe certain basic features, such as the fact that a sphere or a torus has an inside and an outside.

- A. Remember that the Möbius band was an object with just one side.
- B. The object of the mind that we constructed was a combination of a torus and a Möbius band. Does this object have one side or two?
- C. The Klein bottle has a Möbius band in it, so it seems that it has only one side. If we examine the Klein bottle, we see that, indeed, that is the case. If we could trace the Klein bottle with a finger, we would start on what looks to be the outside and end on the inside, never lifting our finger or crossing a boundary.
- D. The Klein bottle is not real and cannot be constructed, but we could build one in the fourth dimension.

V. Can the Klein bottle be constructed from Möbius bands?
- A. If we tried to join the long edges of two Möbius bands together, we would find that as we came around the band, we would be trying to join the edges in opposite directions in the same place.
- B. We know, however, that the Klein bottle is associated with the Möbius band. How is the Möbius band contained in the Klein bottle? How do they fit together?
- C. In mathematics, we always want to work in the domain of the simple and familiar, then apply our experience to the more difficult and unfamiliar.
- D. Let's look at a representation of a familiar shape, the torus; then, we will try to apply our knowledge of it to the Klein bottle.
  1. Again, we start with a construction diagram of the torus. Using a piece of paper, we glue one long edge to the opposite long edge to make a cylinder; then, we glue the two ends of the cylinder together.
  2. Note that a torus is really two cylinders with their ends joined. This knowledge helps us learn the relationship between the representation and the object; in turn, we will apply this understanding to the Klein bottle.
  3. We will cut our construction diagram of a torus with straight lines across, dividing the paper into thirds.

4. Even after we make the first cut, we could still follow the construction diagram to form a torus.
5. If we make the second cut, the result is two cylinders that could be constructed to form a torus.

E. Can we apply the knowledge gained from this experiment with the torus to the more abstract object, the Klein bottle?
1. Is the Klein bottle really two Möbius bands joined together along their edges? Looking at the Klein bottle model does not help us answer this question.
2. We start, again, with a representation, that is, a construction diagram, of the Klein bottle.
3. Again, we cut the construction diagram. This time, what looks like two cut lines is really one because of the way in which the Klein bottle is constructed.
4. Following the points at which the constructed object is to be joined, we see that the center piece resulting from the cut is a Möbius band.
5. We join the remaining two pieces at the points marked earlier and discover, again, that the result is a Möbius band.
6. Our cutting experiment shows that the Klein bottle is really the union of two Möbius bands that are joined together along their edge.

F. This lecture has shown us that we can tame abstract ideas by working with representations of them that allow us to see the concrete.

Suggested Reading:

Edward B. Burger and Michael Starbird, *The Heart of Mathematics: An invitation to effective thinking*, Key College Publishing, Section 5.2, "The Band That Wouldn't Stop Playing: Experimenting with the Möbius Band and Klein Bottle."

Questions to Consider:

1. Suppose you take a square whose edges are identified so as to create a Klein bottle. Suppose you cut the square with three parallel lines going between the twisted sides. How many cuts are you making? How many pieces do you get? What are they?

2. Suppose you take a square whose edges are identified so as to create a torus, that is, the boundary of a doughnut. Draw something on the square such that after you assemble the torus the thing you drew would become one closed curve that goes three times around the torus lengthwise while it goes twice around it widthwise.

Lecture Fourteen—Transcript
A One-Sided, Sealed Surface—The Klein Bottle

In the last lecture, we were introduced to this beautiful figure, the Möbius band. The Möbius, as you recall, has just one surface, one edge, just one side. It's a beautiful object, but not only is it a beautiful object, but it brought up an idea of how we can go about investigating objects, and how we can think about them in effective ways. By looking at an effective representation of the Möbius band, we saw how we can predict what would happen if we cut it in various ways, and we saw that the pre-construction diagram was actually a useful thing to look at when we were trying to explain the whole object.

Well, that concept in itself, the concept of representation of an object, like the Möbius band—a representation of it as a rectangle with edges glued in a certain way—that representation of it turned out to be useful, and then, it is the idea of representing things. Can we represent other, familiar objects in these different kinds of ways, like objects by gluing edges together, in order to gain insight into familiar objects?

One of the strategies of mathematics that is most valuable is to abstract ideas. You have an idea, maybe it's a very concrete idea, but then you take it, and you push it to different domains, and you abstract it to different realms. We will see in today's lecture that those methods, looking at those representations, and then pushing them into the abstract, are going to produce for us things that are not only conceptually beautiful and surprising, but also can make artistic objects that are actually very attractive and have been seen in museums and artworks throughout recent years.

Let's begin, then, again with this concept of the Möbius band, and its representation. Here was a completed Möbius band, but as you recall, it was a very simple construction diagram that made a Möbius band. It was just a rectangle where we showed that we took this end, and we showed how to identify the points on this edge with the points on this edge. Namely, we put a half-twist in them, and those marks on it indicated which points got mapped and glued to which points. Okay?

Well, that's an idea, right? The idea that you can take something that's a surface, something that doesn't sit flat anymore, but you can create it by taking something that is flat, and then showing how to put it together. Let's see if we can apply that to some other familiar surfaces.

Well, the most familiar surface is the sphere. This is a surface that is just completely round, and we want to know if it would be possible to construct a sphere using that same concept, the concept of gluing edges of a piece of paper together to create a sphere.

Well, let's just go ahead and try it. Let's start with a square. This is a square piece of paper, and now, we need to decide what sides are going to be glued to what sides in order to create this potential sphere. Well, let's just go ahead and mark them, and see. If we take this side right here, and we identify this side to this side—we will mark in a different color for this one; we mark this one and identify it to this one—and then we just glue this side to this side, like this, then, that gives us something like an open taco, and then, if we glue this side to this side, then—they are already close to each other—and we can sort of inflate it, we would, in fact, produce a sphere. Right?

You see, it looks almost like a sphere. It doesn't have the roundness to it, but if you inflated it, it would look like a sphere. We have to be a little abstract here. We're not really taking a piece of paper. What we are really taking is a piece of rubber, and then after we've glue this together, and glued this together, and inflated it, if it were made of a sort of balloon material, then it could become a sphere. You see, actually, in my mathematical world, I am a topologist, and to a topologist, anything that can be inflated to look like a sphere is a sphere. To me, then, a football is a sphere, because it could be inflated to look like a sphere, if it were just a kind of rubber thing. It could be made to look like a sphere, so I'm thinking in a more abstract concept of creating these objects.

This is, then, one way to create a sphere, by just gluing these two edges together, and then gluing these two edges together, to get a sphere. We've taken a method of representation that we learned from the Möbius band, and applied it to construct a familiar object, the sphere. Let's try one more of these.

Let's take this torus. Now, a torus looks like the boundary of a bagel, like the boundary of a donut, and it can be constructed by taking a piece of paper, and just doing what you expect, namely, how would you construct something that looks like a torus from a piece of paper? Well, you would first roll it up and make a cylinder, and then you would bring the two ends of the cylinder together, and glue them together to make the torus.

Now, let me indicate exactly how we would do the gluing, because there are choices. Remember that in doing the Möbius band, we had to indicate whether the top edge here went to the bottom, as it does here, versus if when we just went straight—we created a cylinder. We have to be careful, then, and make sure that we specify what points go to which points in the gluing.

Here we go, then. This is the blue side, and we will take this direction of gluing. We'll indicate that this side gets glued to this side, just straight across. Okay? Just straight across. Then, we want this side to get glued to this side, in order to make the torus. Well, look what points get glued to what points. We'll make them in red, here, that the points on this side, in this direction, get glued to the points on this side. There's no twist involved when we create a torus.

We first take it like this, then, roll it up, to make a cylinder, and then roll it up to make a torus. Okay? In your graphic, you will see a neater example of that, to see, in fact, that the arrows are going in the correct direction.

We have seen, now, how to make two familiar objects from our concept of creating a construction diagram of a surface, but now, let's move on to and try something new. One of the ideas of mathematics—and I think it is a terrific idea for trying new concepts, and new ideas, and forcing yourself to think a new thought—is to take a method for doing something familiar. We now have a method for creating a torus, and combining it with some other idea that we have seen before. Well, the idea that we have seen before was to put a twist in before we glued things together. Could we combine these two ideas: The one idea of gluing edges together to create a surface, with the other idea of putting a twist in? In other words, could we, without thinking of what we're going to get, could we take a piece of paper, and say, "I'm going to roll it up to make a cylinder, just as before, but I am going to use some of the idea of the Möbius band to glue these two edges in the opposite direction, as though it were a Möbius band?"

Here we go, then. For this one, I'm going to put the arrows this way and then downward this way. Well, let's just go ahead and try it, and

I'm going to try physically here. We'll just glue this together like this, and that's no problem, to make a cylinder. But now, if we glue it like this, you are going to have a hard time seeing this, so we're going to have a graphic that shows you—you will see now on your screen that if I try to put these things together, the arrows do not match up. They're going the wrong direction.

Now, we might try various ways to get those arrows to match up. For example, we can rotate it to try to match them up, but they just don't match up. The arrows don't match up, because if we glued it just like this, we would get a torus, which had the other identification, the other direction of identification. Since they don't match up, then, and if we twist and they don't match up, we are sort of stuck. What can we do? I mean, we have an idea for constructing something. We have a diagram that tells us what to glue to what, but the only trouble is, we can't do it. The trouble is, if we hold the two edges straight up, then the arrows are going the same direction, but if I turn it, you see that the arrows turn around, so that the clockwise and the counterclockwise switch. Can we get them to glue together when they are both straight up?

Well, the trouble is that you can't, you see, because if you squish them together, that doesn't work, because you aren't keeping them circular, so what we need to do is to think a bit abstractly. How can we actually make the gluing that is indicated by the construction diagram, and make it happen, even though we have trouble physically doing it?

Well, okay. If you have trouble doing something, you have to cheat; you see, that's a very important lesson in life. You have to cheat, so how are we going to cheat? We know that if these two things were just brought up right together, they would fit, because the orientation is correct when they are both going straight up.

Well, one way to do it would be to be take one end, and make it a little smaller. If it could somehow miraculously pass through the sidewall of the other one, and then come up inside, that it would fit perfectly. Of course, though, this passing through the sidewall is really cheating, because there was nothing in our construction diagram that said, "Make a hole here, and push it through," but let's do the best we can. Let's go ahead and do that; make a hole in the sidewall, and then go ahead and push it through. Well, if we do that, we'll get a good object, but it's not exactly what we want. The problem with it is that has this hole in it. We had to pass it through itself in order to construct the object.

Well, here's where the concept of abstraction comes in. We have an object in our minds. In our mind's eye, we can say, "I want to think about that object that I would get if it were possible to glue those ends together in the way that could glue them if I passed through the wall, but if I could do it without passing through the wall."

Now, people who live in the world of mere reality might say, "Well, since you can't do it, how are you going to proceed?" That is no obstacle to a mathematician, though, and no obstacle to people who think abstractly. You see, this is where you get new ideas; it is to go beyond the bounds of mere reality. In this case, therefore, what we're going to say is: Suppose you passed through the wall without touching it, and glued it together? You would get an object of the mind, and that object is a famous object, and it's called the *Klein bottle*, named after a mathematician, Felix Klein. The Klein bottle is a wonderful object, and it is the object you get by making the gluing that cannot happen in reality. Most of it can happen in reality, though, and you can actually construct a model of the Klein bottle, as you see here.

You see, what happened here is that you had this tube, the tube that we got when we glued the top and the bottom, and we made the cylinder. We are thinking of the cylinder as being made of a rubbery kind of substance, and so we glued it back. We needed the end of the cylinder to come to the inside before it glued to the other end of the cylinder, and you see that that is exactly what we did here. You see the cylinder; as I am passing along the cylinder, it goes inside itself, passes through, and then comes back.

Now, this is not actually a Klein bottle, because of the fact that in three-dimensional space, here, where we live, this is passing through itself, and yet the concept of a Klein bottle is that it is just what you would get if you could take this piece of paper without actually intersecting itself, and create an object. It is a beautiful object, nevertheless, in this representation of itself, and it has many properties that we will talk about in a minute.

Okay, so this is a Klein bottle, but there are some features that we associate, or that we observe, in these familiar objects, like the sphere and the torus. A basic feature that we observe is that there's an outside to it and an inside to it. That's what it means to be a closed surface, right? Because if you were an ant crawling around on the outside, there would be no way that you could get to the inside. Likewise, for this torus, if you had the torus, which, remember, is the boundary for this donut, if you were crawling around on the outside, there would be no way to get to the inside.

Now, though, let's think about combining ideas again. In the Möbius band, remember that we had created an object that had just one side. If you were an ant crawling here, and you crawled along and crawled along, you could get to what appeared to be the other side, what we really saw was the same side.

Now, we have constructed an object of the mind, and it was a combination of a torus-like thing and a Möbius band-like object. We can then ask ourselves the question: Does it have only one side, or does it have two sides?

Well, a Möbius band has only one side, and you can see that the Klein bottle, when we look at just this direction of it, has a Möbius band in it, so that it seems like it only has one side. Now, let's look at it, and see that indeed, that is the case. The surface of the mind has only one side. Look, if you come on the outside—here I am, in the outer air, and I come inside, right here, what looks like inside, I come up through this hole, and now here I am, on what looks to us like the inside, the other side of this surface, and I never crossed a boundary. Of course, there was a little bit of cheating, because I had to go through that hole there, but if you thought of the Möbius band as the abstraction, the abstract idea, then there would be no necessity to cross through itself, and you would see that it was a surface that has only one side.

Now, by the way, I've talked about it not being a real object, that the Klein bottle is not real, but that's not really true. In a sense, it is real, and you can actually construct a complete Klein bottle without this hole, and I will tell you that the only thing you need to do is go to the fourth dimension. Because, you see, that hole—you've got most of the Klein bottle here in three dimensions; all you need to do is to take that hole and cap it off into the fourth dimension. That would be a way to actually build the entire Klein bottle without any self-intersections, without any of this crossing through itself, which is not really legal in the world of the fourth dimension. Now, you may quibble and say that the fourth dimension is not very real to you, but I will pretend that I didn't hear that part.

Now, once again, when we are looking at an object, it is often easier to look at a representation of it than to look at the completed object, particularly if the object is a complicated thing than like this Klein bottle, and an abstract idea like the Klein bottle. What I want to do, then, is to ask a question now about whether the Klein bottle can be constructed from Möbius bands. You see, one idea of constructing an object that has just one surface to it would be to take a Möbius band and just start gluing the edges together. You see what I mean? You can just start gluing the edges together. If you did that, though, you would find that when you came all the way around the Möbius band—when you were gluing them on the inside one way—when you came around you would be trying to glue them on the other side the other way, so that it would be impossible to actually create the entire finished product.

This is a problem, then, but nevertheless, we know that a Möbius band is associated with the Klein bottle, because of the twist. The way we created the concept of the Klein bottle was to look at the representation of it as a piece of paper, and then, it had the twist that is associated with the Möbius band.

What I want to investigate now is to look at this Klein bottle, this complicated surface, and ask ourselves how it could be constructed with Möbius bands. How are they in there? How are they fitting together? One of the really valuable lessons from mathematics is that when you are trying to do something that's difficult, the first thing you should do is to quit. I'm really serious. You should just quit. You should never try to do something difficult right away. What you should do is think to yourself, "Is there something easier that is maybe similar, related, that I don't know how to do either, but that is simpler?" Then, that is what you should work on. Work in the domain of the familiar, and then get so strong there that you can apply it to the unfamiliar.

Well, the way we started this discussion is that we looked at a representation of a familiar object, namely, a torus. A torus is very much like the Klein bottle, because it's made with a cylinder, but the gluing works, and let's look at the representation of the torus, and see if we can understand how it is built out of pieces, and then we can go to the representation in the Klein bottle, and try to get the newfound strength that we have from our investigation of the torus, and apply it to the Klein bottle.

Here we go. What we are going to do is look at this representation of the torus, so here is a torus. A torus, as you know, is constructed by taking a piece of paper and gluing the top to the bottom, and the side to the side, with no twists. Okay? I am going to hold this up, and you can see that this is going to be a torus, if I glued the top to the top, and then glued the sides to the sides, as indicated. What I want to show you is that a torus is really taking two cylinders and gluing their ends together. Now, of course, you could see that. If you actually thought of a constructed torus, you could just cut it here and cut here, and then each half is just a cylinder, and then you've glued the top to the top, and the bottom to the bottom, to create this torus.

The reason that we want to do the same process in this somewhat unfamiliar representation, in order to teach us the relationship between the representation and the object, is so that we can apply it to this more abstract thing, the Klein bottle.

Let's go ahead and do this. What we are going to do is this: Simply cut the torus by straight lines across. Here's one here, here's one here. We just cut it straight across, and I am actually going to cut these things, you see? This is the cut. I'm going to cut it straight across. Here we go. I'm first just going to cut one part of it. By the way, notice that this middle thing, after I cut it, this part is to be glued to this, by the construction diagram, and so that would be a cylinder. It would look a little bit like a tuna fish can, that sort of cylinder.

Now, I have cut off this top part, but remember that this picture was a representation of a constructed object, of the torus, and this side of it is indicated to be glued to this side, so that I could actually do the gluing, you see? I could put it down here, and actually tape it up, and that I would have a new representation of the same object, so I am going to go ahead and do that. I'm going to tape it up right here, tape it together, and now, I have a new representation of the torus.

Now, by the way, I made this cut, and it used to be attached to here, so if I want to be clear about how I would reconstruct the whole torus, I had better tell us that this side here is to be glued to this side here. You see? That would be how I would reconstruct it.

Now, if I cut it once again, on this line—see, you can just cut these things up—look what I have. I have two cylinders. This is a cylinder, and this one is a cylinder. You see how this triple arrow goes to this triple arrow, this single arrow goes to this single arrow, so that these are two cylinders? This side gets glued to this end of this cylinder, and the unmarked sides get glued to each other, so that I have used the representation in a way to show you how to take a torus, cut it up into pieces, and make it into two cylinders.

Now, the reason I did that is to try to teach myself something. I was trying to teach myself a way to deal with the more abstract object, the Klein bottle, so let's turn to the Klein bottle representation, and see if we can see that the Klein bottle is really just two Möbius bands sewn together along their edges. You see, each Möbius band has one edge, and they are sewn together to create this Klein bottle, but I defy you to just look at this Klein bottle and see those two Möbius bands in there. It is very difficult to just look at this and say, "Well, where are these bands?" It's very hard to see them, and yes, they are there, as we will see by concentrating on the abstract representation.

Here we go, then, starting again with the representation of a Klein bottle. The way the Klein bottle goes, remember, is that this side is identified to the other side, but in reverse order, with a twist. This is identified to this, but with a twist; and then the top and the bottom are identified without a twist, so that the top is identified to the bottom without a twist, so there it is.

This is a Klein bottle, meaning that it is the construction diagram that would give us a Klein bottle if we actually followed the rules. When I look at this, then, I say to myself that it is the concept of the Klein bottle.

Now, I want to see that this Klein bottle is actually two Möbius bands. Well, I know that the twisting part comes from these sides, so here's what I'm going to do. I'm going to just cut the Klein bottle with the following—I was going to say, and I'm going to ask you to think about whether it would be more appropriate for me to, say, cut the Klein bottle along these lines, or whether it would be more appropriate for me to cut the Klein bottle along this line? This is actually just one line, because, look at this. You see this point here? You march along, march along, march along, and then when you get to this side, what is that point identified with? It is this end of the line, and it continues across, and then this point is identified to this end, so that what looks like two lines is really only one circle, so that we're actually cut along one circle, whereas in the torus case, we actually cut along two circles, two ends of the cylinder.

Now, we go ahead and do the cutting, so here we go. We cut, and we cut, and we cut. Now, after we cut, in order to remember that these two things are really supposed to be identified, we better mark them, so we will mark them. Look at how they are cut. These are various points, so I will put A, B, C, just to indicate that these points are to be identified to each other, to be glued back together again in order to reassemble the figure.

Now what I'm going to do is to continue the cut along this simple closed curve, a curve that is closed up, but there is just one of them; here I go, and when I do this, I now have three pieces of paper, but it is the middle one I'm most interested in right now, because the middle one—do you see what it is? It's a Möbius band. The middle one is a Möbius band, so here I have constructed a Möbius band as one of my pieces. You see? I can actually do that, glue it, and construct one Möbius band, but I've got these two pieces left over. Now, what am I going to do with these two pieces?

Well, do you see these pink arrows? The original construction diagram for the Klein bottle said that the points along this edge were to be glued to the points along this edge. Well, now, since I have them separate, why don't I glue them? I'll go ahead and glue them. Here we go; I glue them together, tape them together like that, and now, look what I have. Do you see what this is? This is a Möbius band. Right? Because the triple arrows here, and the triple arrows going up here are glued together, and this one is glued to this one. If I wanted to construct this together, and follow the rules, I would have constructed another Möbius band.

What I have shown here, then—that by doing this cutting, we have shown that the Klein bottle is actually the union of two Möbius bands that are glued together along their edges. This is sort of a neat feature of the Klein bottle, and I think very difficult to see. If you want to look at the Klein bottle and actually try to see it, here's the way you would do it. You would just take it and cut it exactly down the middle, like this, and then think very hard to see that, in fact, each half of that is actually a Möbius band. It is very difficult to see, though, because it is an image in the mind, very difficult to see.

Using this representation, though, we can conclusively detect that indeed, that is the case.

The Klein bottle occurs in many artistic renditions. This is just one. I recommend that you look for it in museums and in artworks. I think that what we have shown in this lecture is that sometimes, by taking an abstract idea, a concept that is very difficult to get your hands around, you can tame it by looking at a representation of it that brings it down to some concrete issue, and relate it to a familiar object before you start working on the more abstract one. This strategy is very potent in mathematics and can be very potent in other walks of life.

Lecture Fifteen
Ordinary Origami—Creating Beautiful Patterns

Edward B. Burger, Ph.D.

Scope:

In this lecture, we consider the seemingly trivial act of repeatedly folding a sheet of paper. Although such an uninteresting endeavor initially appears to be devoid of intellectual offerings, we will soon discover that nothing can be further from the truth. By examining simple objects deeply, we uncover a treasure trove of hidden nuance and intriguing structure. Suddenly, out of the chaos, patterns emerge, and the at-first unpredictable "randomness" becomes completely understood and intuitive. Using this simple approach, mathematics allows us to journey far beyond the confines of our physical world and extend ourselves outside the physical act of folding paper. Although these patterns and beautiful structure are intriguing in their own right, they foreshadow even greater conquests ahead.

We close this lecture by exploring a dragon-esque fractal and discovering the incredible property that the infinitely jagged shape can be fit together with other copies of itself like a jigsaw puzzle. We could even use this infinitely complicated Dragon Curve to tile a bathroom floor. It would appear that describing the construction of such an infinitely complex structure with the additional property that copies of it can tile the plane would require complicated mathematical theories out of our reach. The lecture that follows this one will reveal a surprising revelation.

Outline

I. We begin this lecture by looking at circumstances that seem to be completely chaotic, but we will uncover hidden patterns and structure in them.

    A. We continually fold a piece of paper in half, then unfold it to reveal what appears to be a chaotic mess. If we look at the unfolded paper, we see nothing more than a series of ridges and valleys.

    B. We can map out these ridges and valleys in a chart showing the paper-folding sequence. Looking at this chart, we see that every sequence begins with two valleys.

- **C.** In fact, if we look at the chart more closely, we see that for each number of folds, the starting sequence is the same as the whole sequence for the previous number of folds. In other words, if we fold the paper four times, then unfold it, we see that the starting sequence is the same as it was when we folded the paper three times. We have identified a pattern in what, at first, seemed chaotic.
- **D.** Using the chart, we can also predict that the first new fold we add after the starting sequence is always a valley.
- **E.** Refolding the paper, we see that each ridge is actually a valley turned upside down. From this observation, we know that the paper-folding sequence after the valley in the middle is an upside-down repetition of the starting sequence. In other words, for each number of folds, we start with the whole previous sequence, add a valley, then flip the previous sequence to determine the current whole sequence.
- **F.** This knowledge enables us to determine the folding sequence for eight folds, even though it is physically impossible to fold a piece of paper eight times.

II. Let's now look at this fold sequence in a completely different way.
- **A.** If we imagine spreading out the folds and inserting alternating ridges and valleys between them, we actually see the next sequence. This pattern has some amazing implications.
- **B.** Let's convert the valleys and ridges to numbers so that we can talk about the sequence in a more precise way. A ridge will be 0, an even number, and a valley will be 1, an odd number.
- **C.** We'll next define a new way to add numbers, in what we call *parity arithmetic*:

    $0 + 0 = 0$ (even number + even number = even number)
    $0 + 1 = 1$ (even number + odd number = odd number)
    $1 + 1 = 0$ (odd number + odd number = even number)

- **D.** We write the original paper-folding sequence as a series of 1s and 0s, then we space out the same sequence underneath and add the two using parity arithmetic. The result is 1000100010001....

E. Mathematicians represent this pattern with a complicated formula called a *power series* that is studied in calculus.
III. We'll close with a subject that seems completely different from what we've been talking about—the fractal known as the *Dragon Curve*.
   A. A fractal is an object that is infinitely complicated. If we look closely at any portion of a fractal, we see yet more complexity.
   B. If we make copies of the Dragon Curve, we can fit them together like a jigsaw puzzle, even though they are infinitely intricate. In fact, we can fit copies of the Dragon Curve together in such a way as to tile the plane.
   C. How do we generate such symmetric complexity? As we'll see in the next lecture, the answer to that question is related to paper-folding.

Suggested Reading:

Edward B. Burger and Michael Starbird, *The Heart of Mathematics: An invitation to effective thinking*, Key College Publishing, Chapter 6, "Chaos and Fractals."

Questions to Consider:

1. Take a piece of paper and fold it repeatedly left over right rather than right over left as was done in the lecture. Write down the first four sequences of valleys and ridges. How does the pattern created by right over left folds compare with the pattern of left over right folds?
2. In going from two folds to three folds and three folds to four folds, explain using actual paper why the method of interpolating alternating valleys and ridges between the folds at one stage produces the sequence of folds at the next stage.

Lecture Fifteen—Transcript
Ordinary Origami—Creating Beautiful Patterns

Well, in the next two lectures, we're going to move from Mike's wonderful discussion of the twisted world of the Möbius band and the Klein bottle to a whole new realm of mathematical vistas. This realm, which we think of as journeying from just folding paper, all the way to the infinite beauty of fractals, is one that's filled with intrigue. And the first two lectures that I'm going to present here in this section are really lectures that set the stage for what Mike will talk about in the next two lectures, when he talks about the wonderful ideas of fractals, which we'll mention in a minute.

Really, when you think about the lectures that we've developed up to this point, we have often seen in this class that very complex phenomena can be understood by looking at a simple circumstance. We have seen this lesson time and time again, and the import of it cannot be overemphasized.

Well, here, in the next two lectures, we're going to explore the consequences of simple, repeated processes, and look at those things closely, find a pattern, and see that by repeating something again and again and again, some amazing structure appears. Alexander Pope once wrote:

> Not chaoslike together, crushed and bruised,
>
> But, as the world, harmoniously confused;
>
> Where order in variety we see,
>
> And where, though all things differ, all agree.

That really is the point of this lecture. We are going to begin by looking at circumstances that seem completely chaotic, and without any kind of coherence or structure, and from there, we're going to uncover invisible patterns and deep connections with nature.

It's all going to actually happen out of a harmless piece of paper. We start with a piece of paper, then, and we're going to produce a very, very simple operation. In fact, you can think of this as a sort of origami for the origamically-challenged. I know that that's very p.c. He can't say "people who can't do origami." You have to say something that is kind of p.c., so the "origamically challenged."

I don't know if you have seen these things. They have these figures; people can make these swans, and they are just amazing. I don't know how to do that, but I can do this, and I'm going to teach to you, too.

All we're going to do is take a piece of paper, and fold it in half. It is the simplest thing you can possibly imagine, and you now have a folded sheet of paper. Now, all I'm going to do is repeat that, so I'm going to take this, and fold again, right over left as I look at it. I guess from your vantage point, it's left over right, but I'm going to continue to take right over left, and fold it again, and again, and again.

Now, in the very first real lecture, Lecture Two, Mike actually talked about folding paper, and how thick the paper gets. In fact, a great little trivia question is: How many times can you physically fold a regular sheet of paper? It turns out that the answer, I've heard, is seven, but I have yet to fold it six times myself. Apparently, though, the record is seven folds. If you remember from Mike's earlier lecture, he showed us that, in fact, if you could actually physically fold a piece of paper in half again, and again, and again, ten times, that in fact, the thickness would be three inches thick, which is almost about two reams of paper. If you just folded a piece of paper in half 40 times, in fact, the thickness of the paper would be so thick that it would reach from here to the moon, and if you were actually able to fold the paper in half again, and again, and again 51 times, the paper thickness would start here, and actually go far past the sun. The power of exponential growth.

Well, that's not part of the lecture. All we're going to do here is fold it as far as I can, which I think is either five or six times, but I was not counting, and then what we are going to do is unfold it. You see, Mike never did that. He just folded it, but now here, we're actually going to unfold the paper, and see what we get.

When you unfold this piece of paper, what you see is a sort of jumbled, chaotic mess, and it seems completely devoid of structure or of any kind of pattern, because it's just a bunch of wrinkles. In fact, though, if you look at this really, really closely, you'll see that the wrinkles come in two flavors. Let me actually point out the two flavors of wrinkles. I'm going to use a thinner piece of paper so that we can actually see the folds a little bit more easily.

I'm just going to do the same process, though, so if you want to join in, or try it a little bit later, I'm just taking the right edge, folding it over the left edge, and nothing more, just always right over left, if you are actually going to try this. Let me fold it one more time, and let's take a look at what we have.

When we open it up, what we see is this wrinkled mess. I'll put it up here, so we can all look at it together, and if you look at that wrinkled mess, what you see, in fact, are two basic types of wrinkles, two types of folds. We see these sort of downward type of folds, and then these sort of upward, spiky type of folds, and I'm going to give each of these a name so that we can refer to them. I'm going to call these folds down here "valleys," as though we're in a valley. They are sort of folding, pointing down; and these I am going to call "ridges," so that I see valleys and ridges, and valleys and ridges, and valleys and ridges, and so forth. In fact, in this particular configuration, what we see here is "valley, valley, ridge, valley, valley, ridge, ridge." So that would be the sequence, if you will, if I just list it. With all the ridges and valleys that we would see, we would see that sequence of runs of valleys and ridges.

Well, it seems pretty chaotic. Here, you could read it off too, if you wanted to. It would be really hard, but you could try. I see all sorts of valleys and ridges, and I don't know if you can see that or not. I won't even read it to you, because it's too complicated, but you could read off just looking at each fold and seeing if it is a valley or ridge.

Well, it seems as though there's no discernible pattern here at all. Let me just try to fold two times. Now, if I just fold it once, by the way—this is sort of a complicated pattern; I'm going to slide it all the way down here—that's actually at a pretty easy pattern, and actually, a nice origami, if you like origami. It's a flying bird. Do you see that? Isn't it beautiful? If you want to impress your friends, again, origami for the origamically-challenged, there is the flying bird, which you can see, is very simply, just that valley, so that's easy.

Now, let's try two folds. Now, the power of looking at simple things deeply really cannot be de-emphasized. I just folded it twice. Let's open it up, and see what we get. Let's put underneath it, so we can look at it, and what we see here now is "valley, valley, ridge." Okay? That seems fine.

I'm going to do it one more time, and I think that's actually that image right here, and we see "valley, valley, ridge, valley, valley, ridge, ridge." Well, there seems to be nothing happening here at all except that the figures are getting more and more complicated.

Let's actually list these things, though. Take a look at this image, and you see that—I call this the "paper-folding sequence," because it's just the sequence of valleys and ridges that we see, and you can see that when we start off, we just have that valley, that flying bird, and the next sequence we have here is "valley, valley, ridge." Then, underneath it, I write the one that we see after we fold three times. We see "valley, valley, ridge, valley, valley, ridge, ridge."

Now, if you were to fold it four times, what we would see is "valley, valley, ridge, valley, valley, ridge, ridge, valley, valley, valley, ridge, ridge, valley, ridge, ridge." Okay, now, we didn't even fold it four times, but if you fold it four times, in fact, that's what we would see.

Look at the chart up to this point, though. What do you notice? What you notice is that when you look at the chart in this fashion, there is some structure that becomes evident. For one thing, you notice that everything begins with "valley," so we always start with a valley by this folding process, even the more complicated ones toward the end.

That's interesting; in fact, if you look closer, you see that, in fact, we always start with "valley, valley." In fact, once you see that, you begin to look at the table and realize that if you look at any one level, the beginning of that level is precisely the previous level. Look, for example, at the third level down. We see "valley, valley, ridge, valley, valley ridge, ridge." Well, look at the first three folds. It's "valley, valley, ridge"—mainly, the folding sequence of the previous thing, so that we are seeing some kind of pattern here, and in fact, let's see if that continues.

Let's look from the three-folds to the four-folds. What happens? Notice that they all do agree. If you look at the four-folding sequence, we see "valley, valley, ridge, valley, valley, ridge, ridge." That beginning is precisely the same as the entire run of the three-fold. Somehow, then, when we fold a paper three times, the beginning of that, that sequence right there, gives us the beginning of the four-fold. If we were now to fold it four times, four times right now, that's one, two, three, and now, I fold it one more time, four, and we open it up, we do indeed see that the first few people are precisely this sequence. I will put it on here, and let's just take a look. "Valley, valley, ridge, valley, valley, ridge, ridge." That completely corresponds to the thing after the "ridge, ridge."

In fact, therefore, there is structure here. Right? The structure is, in fact, that if you want to know how the next fold sequence would begin, what would be your case? Your guess is that the first half of that next folding sequence would be precisely this sequence of folds right here, and then some more stuff, yet to be determined.

All of a sudden, notice that once we look at this carefully, we see patterns. Now, once we see structure, suddenly, other patterns come into focus. Now, what I want to do is see if we can figure out what that ending pattern is. Well, let's take a look at our chart again.

If we look at our chart, we see, for example, if we look at the second level, that the first person, of course, is the first fold of the previous one, just that valley, and then, we have another copy of a valley, and then we have something else.

Let's see what happens when we look after three folds. After three folds, we begin with the previous "valley, valley, ridge." Then, notice that we again have a "valley," and then some stuff.

Let's look at four folds. I see the previous "valley, valley, ridge, valley, valley, ridge, ridge," and then, notice the next one is, indeed, a "valley." It always seems as though, after we run the course of the previous folds, we introduce a "valley," and then some other stuff. There's another great observation. The first new person we add, in fact, is always a "valley." If you want to figure out what the folding sequence would be if you fold one, two, three, four, five times, the answer is that we would start with this run of "valleys" and "ridges," and then, we would add a new "valley," and then some other stuff.

Now, what about the other stuff? Well, look at the images again, at the chart, and we see something utterly fantastic. If we look at the two-fold's thing, we see "valley, valley, ridge." Well, we understand the "valley." That's from the previous thing. We understand this new "valley," because that's that center valley we add, and now, we've got a "ridge," and notice that a "ridge" is just a "valley" on its side. In fact, all we did was to basically take the first piece, that little "valley," and turn it upside down. In fact, if you look at the physical piece of paper, you can actually see that, because if you look at this sheet of paper right here, notice that when I turn it back, they're both "valleys" now, and yet, when I open it up, since I'm opening it, that "valley" becomes a "ridge."

Well, once you make that very simple observation, the pattern becomes completely clear. Let's look at our chart one last time, and now, watch the structure unfold. Let's look at what happens after three folds. We have "valley, valley, ridge." Then, we have that center "V"—"valley"—and then, what comes next? It's just the previous people that we had, but flipped over, unfolded. Read the beginning backwards and upside down, so start from left to right at the beginning half, and you see that that "ridge" becomes a "valley," and those first two "valleys" become "ridges." It literally is a flip, then, and you can see that in the animation; how, in fact, we can flip one to match the other.

Let's look at four folds, where it's even harder to say, but you see that we have, first, the run of three folds, then that center "V," and notice that what remains, if you were to fold it back onto the first half, matches up perfectly.

What you do, then, if you actually want to produce the next level, if you fold one, two, three, four, five times? Well, if you were to fold it five times, what do you do? You copy all of the "valley, valley, ridge, valley, valley, ridge, ridge" sequence here, you add on that center "V," and then, what do you do? You take this, and you flip it upside down, and continue reading it off, because we keep folding it in half.

Well, once we are armed with this pattern—like a domino effect, all of these want to fall down—we can now continue the process. For example, what would the folding sequence be for six folds? Well, it's easy to do. You write down what you have for five, put the center "V," and then take your answer, and flip it over, and keep reading.

What about seven folds? The maximum that we can humanly fold actual pieces of paper to? No problem. Put down the run of six folds, and do what? Introduce a "V," and then take this sequence, and flip it over, and continue reading.

The great thing is, though, that once we have this, we can actually transcend the physical world. What would the folding sequence be if we could actually fold a piece of paper eight times? We can't. We physically cannot fold a paper eight times, but if we could, we can actually say what the sequence would be. It is just the run of when we folded it seven times, introduce that center "valley," take the sequence, flip it, and continue reading. Thus, you could even do, and tell, the precise sequence of folds for 43 folds, which will take you all the way to the moon, or 52 folds, which will take you past the sun. You can actually write down that sequence of valleys and ridges that take you all the way out to outer space, and we can do that because we found this pattern.

Now that we have this amazing pattern, all of a sudden, we can see even more surprising structure hidden inside. Remember, the important thing here from my vantage point is that when we started, we just had this randomness, this jumble, and now, the jumble has been tamed by just looking at these simple things deeply, and finding a very nice pattern.

Now that we have patterns, let's look for more. In particular, let's look at this fold sequence in a completely different way. What I want to do is just, now, take the first valley that we have, the first case, and all I am going to do is to put down a very simple sequence. I'm going to flank this with just an up-down move, so just alternate between up and down. Notice that what I do that, I actually get the next sequence. Do you see that? Because that's just that center fold, there's the center fold, and then I just have an "up-down."

Well, what if I repeat that process? What if I take this, and now just add "up-downs" in between everything? Well, take a look at this chart here, and see what happens. If I spread out the "valley, valley, ridge," and in between them, I just insert an alternating sequence of "valley" and "ridge," "valley, ridge," "valley, ridge," "valley, ridge," what do I see? I actually see "valley, valley, ridge, valley, valley, ridge, ridge." I actually see the next turn in the sequence, and this process continues.

If I now spread out that so that I can insert things in between, and I just put an alternating sequence of "valley, ridge, valley, ridge, valley, ridge" in between, then if you read out the totality, what do you see? You see "valley, valley, ridge, valley, valley, ridge, ridge, valley, valley, valley, ridge, ridge, valley, ridge, ridge." You actually see the next fold sequence.

That points out an amazing hidden piece of structure. If you look at every other person, every other fold, you see an alternating sequence:

"valley, ridge," "valley, ridge," "valley, ridge," "valley, ridge." It alternates. Again, we start with chaos and see pattern.

Now, this particular pattern, in fact, has some very powerful implications, and I wanted to actually share one of them with you that actually tames this incredible sequence even more.

First of all, let's become a little bit more quantitative, in a way, and what I'd like to do is to first convert these valleys and ridges to numbers, so that we can actually talk about this sequence in a more precise way. What I'm going to do is consider any ridge that we see a "0," and any valley that we see a "1." I can then convert these valleys and ridges that we see all over the place into just a bunch of zeros and ones. For example, this sequence here, would just be a "1, 1, 0" and so forth. You could convert the runs of valleys and ridges to zeros and ones.

The reason why this is valuable is because now, if we have zeros and ones, we can combine them. We can actually introduce some very, very simple arithmetic. In fact, arithmetic that is similar to what computers use, which is sort of fun, and here is the arithmetic. I call it *parity arithmetic*, because all I care about is whether things are even or odd, so I'm going to define a new way to add numbers, and all we're going to add are zeros and ones.

Here's the recipe: 0 plus 0 will equal 0. No big surprise. 0 plus 1 will equal 1. Again, you're probably not shocked, but here's the big surprise. If you take 1 and add it to 1, I'm going to declare that, in this new addition, to be 0. That might sound a little bit strange, but if you think about 1 as being an odd number, then, if you think about it, if you take an odd number and add it to an odd number, that result is actually an even number, so that I will think of 0 as kind of representing "even," and 1 representing "odd," and then, the addition begins to make sense. An even plus an even (0+0) is 0. An even plus an odd (0+1) is odd, 1. And if we take an odd plus an odd (1+1), we actually get an even, which I'm going to now declare to be 0. It's kind of a naive sense of arithmetic, but, in fact, it's an arithmetic nonetheless.

Now, let's try the following incredible experiment. Let's write down the paper-folding sequence, a long run of these zeros and ones. Of course, we know that it starts off "1, 1, 0, 1, 1, 0, 0, 1, 1,1, 0, 0, 1, 0, 0," and it goes on, and we know how to understand the patterns, and so forth. Let's write another copy of it down, but let's actually space out that copy so that I write the new version "1 under every other term," so that I'm going to write out the same sequence, "1, 1, 0, 1, 1, 0, 0, 1, 1, 1, 0, 0," just as we had before, but I'm going to space them out, and you can see how I space them out underneath.

Now, what I would like to do is add these two long lists of numbers, but I am going to use this naive sense of arithmetic. What happens when we add them? Well, if there's nothing underneath, like the first one, there's nothing underneath, so that's an invisible 0, so we have 1. Now, though, notice under the next column, we have 1 plus 1, which by our new arithmetic, is, in fact, 0. Then, I have 0 over nothing, so that's just 0. 1 plus 1 is 0; I have 1 plus nothing is 1; 0 plus 0 is 0; 0 plus nothing is 0; 1 plus 1 is 0; and when we fill in all the addition, look at the answer. The answer has this repeating pattern. It's just "1, 0, 0, 0, 1, 0, 0, 0, 1, 0, 0, 0, 1."

There must be some amazing hidden structure in this chaos where absolutely no repetition is appearing just by taking this, associating it with numbers, and adding them in a special way. Well, in fact, you can actually give a very fancy math way of looking at this, and I thought that you might enjoy just seeing on the screen what the math reflections of this would be. This is absolutely just for fun, by the way, not really for any kind of major life lesson, but to show you that, in fact, this pattern allows mathematicians to explore the complicated structure that we see here in a deeper way.

If we actually write the letter "f" for the individual terms in this folding sequence—for example, the "1, 1, 0, 1, 1," would represent the first "f." The second "f," the third "f," and I will call them "f_n," so that f_1 is the first one, f_2 will be the second one, and so forth, then, in fact, you can actually form what is called a generating function. This is actually a very fancy way of looking at all those numbers at once, and you can see it on your screen. It looks like this cryptic code where I have this funny Greek symbol, Σ, which actually represents a "sum," and in fact, this is really an infinite sum. I add up all of those terms, but you'll notice that I have inserted a variable, x, in between each one of them, and the x has a higher and higher power. This is actually an example of an infinite sum, and because I have infinitely many terms, I am assuming that I am folding this forever, but in reality what is really is is an example of what's called a *power series*. This is something that people study, for example, in calculus.

In any case, the interesting thing here is that you will notice that this power series, even though it looks foreign, has in it, in terms of all the values that become before the x's, this sequence. It is all the zeros and ones, in the right order, to actually produce this very scary, impressively foreign-looking symbol. However, the fact we just discovered, this repeating pattern, can be actually written in terms of this function, and the way to express it is the following: If we take that function, F of x, and we add F of x^2, then that equals x divided by 1 minus x^4.

The point here is that the x^2 part is the shifting "1 every other one," and we add the two as before, and we got that repeating thing, which means that instead of an infinitely long thing, we can just express the answer as that little fraction, x over 1 minus x^4. In fact, then, that long list of adding, where we took chaos, spread it out, added it and got something that repeated, we see reflections of it in the math version by taking this impressive formula, and then putting the formula with x^2—that spreads it out—and then we add, and then the repeating thing comes out to resolve in the x divided by 1 minus x^4. I thought it was sort of interesting to see, in fact, the math reflections of that.

In any case, we see incredible structure, and much of it, in two different ways. First of all, we see how to build this run of sequences by just taking the previous answer, writing it down, inserting an additional, new center "valley," and then taking the old answer, flipping it, reversing it, and continue reading. That's one way to produce this long list of paper-folding sequences.

The other way, of course, is to take the previous answer, and then insert in between each the alternating sequence of "valley, ridge, valley, ridge, valley, ridge," and it turns out that we saw that that also is a model for this amazing chaos. Well, I actually want to step away and take a momentary intermission. With all this paper-folding, in fact, you may be folded out, so I want to actually offer you something that seems, and is, on the onset, absolutely, completely divorced from what we're talking about.

Now, then, this is going to be a dramatic intellectual shift for a second. Think of it as a mathematical intermission, if you will, to take our minds off all those funny symbols that we saw. I want to actually show you a picture of something that I think is very beautiful. What you are looking at here is an example of what's called a *Dragon Curve*. It is sort of fiery; if you look at the picture, it sort of has this fierce dragon-looking image to it.

Well, it actually is an example of what's called a *fractal*. Now, a fractal—what we are thinking of, in this course, by the way, is sort of "The Joy of Thinking Through Classical Mathematics." The fractal idea is actually fairly new, certainly within the past 100 years, so maybe that's not too classical, but if we think about classical as being anything before we were alive, then, for most of us, this would even be considered classical.

In any case, what is a fractal? A fractal really is an object that is, in some sense, infinitely complicated, and that is why you can almost see reflections of this idea here in the folds of the piece of paper. You know, we see this object, and if we look closer any portion of this Dragon Curve, we see more bumps and more wrinkles, and the closer we look, the more wrinkles we see. It turns out that defining a fractal mathematically with great precision is very difficult. It is sort of an ill-defined mathematical term, but the idea that it brings to mind is, in fact something that's sort of infinitely complicated. It has creases, on top of creases, on top of creases. The closer you look, the more creases you see, and in Mike's next two lectures after my next one, you are going to see many, many examples of fractals, not only mathematical ones, but also ones that even appear in nature.

I just want us to look at this Dragon Curve, though, because it's so interesting, so beautiful, and so infinitely complicated, and I will show you a neat little fact about this Dragon Curve. The interesting fact is that if you take various copies of the Dragon Curve, just take the Dragon Curve and look at various copies of it, you can fit the Dragon Curve together like a jigsaw puzzle. They are infinitely intricate, so they have infinitely many bumps upon bumps upon bumps, and yet, they fit together perfectly. In fact, as you can see in this sequence of images here, sometimes I might have to turn one of the images completely around 180 degrees, turn it upside down, but if I turn it upside down, it fits back onto itself, and clicks like a jigsaw puzzle, and you'll notice that it can actually be used to tile the entire plane, so that with this infinite complexity, they fit together so perfectly that they cover all of the floor.

The next time you're about to tile the bathroom, you might want to consider this as an interesting alternative to the usual square tiles, because, in fact, this could be used to tile your bathroom floor. Of course, the tiles would be infinitely complicated, and it would probably cost you an arm and a leg, so I'm not actually recommending this, but it's interesting to see that this one infinitely complicated image, the Dragon Curve, an example of a fractal, can be used just itself to cover the floor, to cover the plane.

Now, you're saying, "Okay, Ed, that seems like an interesting little piece of trivia. It seems pretty complicated." Some of you might be wondering, "How do you actually describe or build the Dragon Curve?"

Well, it seems very complicated. If the thing is infinitely complex, that somehow, the method of describing it might have very complicated formulas. It might be certainly beyond my capabilities, potentially. Well, that remains an open question, which I would actually like to hold for the next lecture, but I will foreshadow it by telling you that somehow, that Dragon Curve, with her infinite complexity, is somehow connected to the folds within the paper that we see here, and this might seem almost scary, because, gee, you don't really want to get paper too close to a Dragon Curve, because it could go up in flames, potentially. It's okay, though. The fiery image of the Dragon Curve can be, in fact, tamed, and we're going to tame it just through folding paper.

What we're also going to see in the next lecture is that this paper-folding sequence can be looked at and a variety of surprisingly new ways. In particular, if you think about it, in all the versions of generating the sequence of paper-folding, we always had to use the previous folds in order to generate the next generation. Either we take the old version, introduce that version, and flip it over to get the next version, or we take the old version, spread it out, and introduce those alternating "valley, ridge, valley, ridge." In both cases, though, to get the next sequence, we had to use the previous one.

Well, I leave you with the challenge of seeing whether there's a way to potentially describe this paper-folding sequence without ever mentioning the previous ones. It turns out that the answer is "yes," and somehow, magically, this is going to be connected with the notion of computing, and we're actually going to see that somehow, what the foundational issues are in computing and in programming turn out to have implications to this paper-folding.

In the next lecture, then, we're going to return to the paper-folding sequence, which we now see is not nearly as chaotic as we first thought, and we're going to see reflections of it not only through the dawn of computing, but also in the eyes of the fiery dragon. What's the point, the meta-life lesson point of this lecture? It really is that looking at issues in different ways often allows us to uncover and find hidden structure. If we look for patterns, we will find them.

Lecture Sixteen
Unfolding Paper to Reveal a Fiery Fractal

Edward B. Burger, Ph.D.

Scope:

Leaving the Dragon Curve alone for the moment, we return to an exploration into the paper-folding sequence introduced in the previous lecture. By merely looking closely and carefully at an object, suddenly, rich structure comes into focus. We will see this maxim beautifully illustrated as we uncover and expose many mysteries hidden in the folds of a sheet of paper. The patterns that emerge appear endless; using these mathematical patterns, we are able to go to where no paper folder has ever gone before.

Although these activities may appear to be just for those who are "origamically challenged," we will discover that the paper-folding sequence offers an example of the classical computational theory of *automata*, developed by Alan Turing—the father of modern computing. In a surprising turn of events, we will then close with the realization that the simple paper-folding process is the key to unleashing the secrets of the Dragon Curve. Thus, we journey from paper folding to automata theory and to fractals by just following the thread of an idea and looking at simple things deeply.

Outline

I. In this lecture, we return to the paper-folding exercise and see many different ideas of mathematics come together.
  A. In the last lecture, we saw different ways to identify and predict a simple paper-folding sequence.
  B. With both the "ridge and valley" and the numerical notations, we needed the previous paper-folding sequence to generate the next sequence. Can we find a way to generate the next sequence without knowing the previous one?

II. To answer this question, we must turn our attention to computing and the work of the British mathematician and computer scientist Alan Turing.
  A. Turing introduced the idea of computing in 1932 and, in 1950, developed the first digital computer, the Universal Turing Machine.

B. We will focus on a special example of Turing Machines known as *finite-state automata*, the simplest type of computer imaginable.
  1. A Turing Machine can only read and write. For example, it may be programmed to read a list of numbers. When it reads a 3 at the beginning of a list, it is programmed to write a 6 at the end of the list.
  2. A program for a Turing Machine, then, is really not very complicated; it is simply a list of rules.

C. In another example, we see a program with the following rules: When the computer reads a 1, it should write 2, 1, 2. When it reads a 2, it writes 1, 3, 0. When it reads a 3, it writes 0.
  1. In running this program, the computer ultimately reaches the number 0 and terminates because it has no rules for that number.
  2. This example illustrates the *halting problem* in computer programming.

D. In a third example, we have the following set of rules: If the computer reads a 0, it writes 0, 1; if it reads a 1, it writes 3, 2; if it reads a 2, it writes 4, 2; if it reads a 3, it writes 3, 1; and if it reads a 4, it writes 4, 1. This is an example of a Turing Machine program that will not terminate.

E. As simple as this program may seem, it is still too complicated for our purposes. We'll simplify the program, then, by asking the computer to record only whether the numbers are even or odd.
  1. If the computer reads an even number, it writes a 0. If it reads an odd number, it writes a 1.
  2. The output of this Turing Machine program brings us back to the paper-folding sequence.
  3. Notice that with this output, we do not need the previous paper-folding sequence to generate the next sequence in the series.

III. Let's look again at the folded papers and stipulate that every crease forms a 90-degree angle.
  A. In other words, we look at each of the folded papers as a collection of right angles.

B. Recall the earlier generating process we discussed when we spread out the folds and placed alternating ridges and valleys between them. This process allowed us to generate the next term in the sequence.

C. We can use the same process in this case, but now, the alternating sequence changes from up-down-up (or ridge-valley-ridge) to right-left-right.

D. Again, by weaving through the right angles in the folded papers, we can generate the next sequence of right angles in the series.

E. Michael Crichton uses this right-angle sequence progressively throughout his book *Jurassic Park* to reveal the fractal Dragon Curve.

F. Our quest for understanding has taken us on a journey through paper folding, to pattern recognition, to naïve arithmetic, to Turing Machines, and to fractals. We gained understanding of these realms by looking at simple things deeply.

Suggested Reading:

Barnsley, Michael. *Fractals Everywhere,* Academic Press, San Diego, 2000.

Burger, Edward B. and Michael Starbird, *The Heart of Mathematics: An invitation to effective thinking*, Key College Publishing, Chapter 6, "Chaos and Fractals."

Questions to Consider:

1. Consider the following Turing Machine:

    $0 \rightarrow 01$

    $1 \rightarrow 10$

    If we start with 1, what would be the first 20 digits outputted?

2. Will the program described in Question 1 halt?

Lecture Sixteen—Transcript
Unfolding Paper to Reveal a Fiery Fractal

Well, in the last lecture, we saw the powerful lesson of searching for a pattern as a means to discover connections and develop understanding, and in this lecture I want us to return back to the paper-folding issues, and see some truly wonderful things. Mathematically, we're going to see how disparate and desperate ideas of mathematics come together in a culminating way, which, by the way, is one of the wonderful things about mathematics. This course, which is really "The Joy of Thinking through Classical Mathematical Ideas," focuses on the thinking aspects that we can use in our everyday lives, to make our lives richer, and to resolve issues in a more powerful way.

Both Mike and I really, truly, love mathematics, though, mathematics in the abstract. And one of the things that is so beautiful about mathematics in the abstract is that so often, we see different areas of mathematics as little pockets, unrelated, but it turns out that the reality is that these apparently different-looking things turn out to actually be connected, sometimes very complicated connections, sometimes very subtle connections, but always connected. Why I'm excited about this particular lecture is that we will see many, many different ideas come together, all through the simple act of folding paper.

Well, let's begin. Let me remind you from the last lecture that we just took a piece of paper, and we started folding it, right over left, and we kept repeating that fold, again and again. Now, physically, we could only do it six or seven times, but mathematically, we could think of folding it as many times as we wanted. After doing that a few times, we started to develop this crinkled, chaotic mess that we saw, and then we tamed that chaotic mess by seeing that there was a pattern there, and that pattern, which can be viewed by just looking at these simple cases and seeing the pattern, could see the pattern in a variety of different ways. Let me just briefly remind you of the ways we saw it.

One way to develop this paper-folding pattern is to first take the first figure we get, which is just a "valley," and then merely report that "valley" again, add a center "valley," and then, with the folding method, just replace this with its reverse. Similarly, to get the next one, we just start with this pattern, "valley, valley, ridge," and here is the "valley, valley, ridge," add that center "valley," and then repeat this pattern flipped over, so backwards, and reverse, and report the "valley, ridge, ridge." Similarly, we can then report this in order, add that additional "valley," and then, take this and flip it over, and keep reading to get the next one, and we can repeat this *ad infinitum*.

The other way we saw that we could actually generate the sequence was by starting with the center valley, and then flanking it with the very simple alternating sequence of "valley, ridge." Here, you'll see, "valley, ridge," and those two pieces that are alternating straddle in between the center "valley" to produce the next run. If we take this run, and add in, "valley, ridge, valley, ridge," then, in fact, we get the next run, and let me point out to you, that we get the "valley, ridge, valley, ridge," as the alternating one.

We could then take this, and between each one, add "valley, ridge, valley, ridge, valley, ridge," and alternate. You could almost imagine serpentining through with valleys and ridges in between to get the next one. In fact, this serpentining through is not a bad image to keep in the back of our heads for the very end of the lecture.

In any case, we saw two basic ways to generate this wonderful sequence of folding papers, and we saw that, in fact, what looked chaotic actually could be completely understood, tamed, explored, and even with mathematical significance, as we saw at the end of the last lecture.

Here, I want to look at the sequence itself, and in particular, I want us to return to this sequence as a means of uncovering pattern and structure. Now, what we did at the end of the last lecture was to denote the "valleys" by ones and the "ridges" by zeros. Thus, these runs of valleys and ridges could be expressed numerically in terms of zeros and ones. For example, if you do the paper-folding sequence a lot, just keep going, we always know it's going to start the same way. And in fact, what you would get would be "valley, valley, ridge, valley, valley, ridge, ridge," and so forth. It gets converted into numbers as "1, 1, 0, 1, 1, 0, 0, 1, 1, 1, 0, 0, 1, 0, 0, 1, 1, 1, 0, 1, 1, 0," and so on, a bunch of zeros and ones, but just perfectly placed so that in fact, everything works.

In both the generating processes that we discovered in generating these, the alternating one and then the folding one—we do notice that to get the next person on the list, the next run of paper-folding sequences, it required us to know the previous one, either taking the previous one, adding the center "V," flipping, or taking the previous one, spreading it out to have room to serpentine through with the valleys and ridges alternating. In any case, we need the previous one in order to generate the first one, not unlike the Fibonacci numbers that we saw some lectures back in number theory, where we saw together, that in fact, you need to take the two previous ones, because there, you need two previous ones, and add them together to get the next one. Here, you just need the one previous one.

Well, this raises the question: Is there a way of doing it without actually using the previous ones? Specifically, is there a way of producing the paper-folding sequence of zeros and ones without actually starting off with the previous one to get the next one? Well, that's the question I want us to think about, and if, in fact, the answer is "yes," that would mean that there would be some new additional structure that we have yet to uncover about this now incredibly interesting series of folds in the paper.

Well, actually, to answer this question, we have to turn our attention to the idea of computing, not necessarily the notion of computers, which we use all the time to take money out of the ATM, or to surf the Web. That's an important use of it, but now I'm thinking about the actual programming of computing, and, in fact, it was a great British mathematician and computer scientist, Alan Turing, who first introduced the concept of a computer back in 1932. His idea actually led to the first digital computer back in 1950, which was known as the Universal Turing Machine.

Now, while Turing is known for many important scientific contributions, such as his brilliant insight into deciphering the enigma code, the code used by the German armed forces in World War II in radio communications. Here, we're going to focus on a very special example of a Turing Machine known as a *finite-state automata*. Now, this sounds really scary, but what we are actually going to talk about is the simplest type of computer you can imagine, far simpler than the laptop computer you have on your desk, or the computer you have your office.

An automata is a program, is a computing machine that can actually read programs, and it only does two things, so it really can't be very difficult at all. All it can do is read and write. Let me actually show you what a finite-state automata or a Turing Machine is all about.

This actually is an example of what some input and output would look like.

A Turing Machine does two things. It can read something, so it could read that that number is a 3, and then it can write something at the very end of the list. Now, what we're going to produce here, then, is a list of numbers. We're going to start off with something at the beginning, and then, what a Turing Machine does is put its little reader head right here at the beginning, and it reads that number, and then based on that number, it produces a number at the very end of the list, and it writes a number down. Then, it moves over one, and then, reads the next number, and based on the value of that number, it produces a number in the next spot, so that it is reading, and then, in turn, writing, and thus, we read, write, both read and write. For example, let me just do an example right here. If it reads the "3" here, then maybe the computer program is, "When you see a 3, write a 6." Well, in that case, when it sees 3, it produces number 6 here. Then, when the reader moves over, it sees a 1, and maybe, this computer program is of the kind that says, "When you see a 1, write a 4." Well, it sees a 1, so at the very end of the list, it produces a 4. So now the writer shifts over, and the reader shifts over, it reads a 5, and maybe this computer program says, "When you see a 5, write a 0." Okay, we see a 5, so at the end of our list, we write a 0, and then we continue in this process. We put something at the end of the list, and we keep going.

A computer program in this case, then, actually isn't that complicated. All it is is a list of rules. When you see a 3, mark 6. When you see a 1, mark 4. When you see a 5, mark 0. Those commands are actually a computer program, and so this is the output of a computer. Now, you're saying, "Gee, does that help balance my checkbook?" Well, it turns out that the answer is that it can be used to balance your check if you write the elaborate computing code required. This is a very simple code, but, in fact, you can write a more elaborate code.

The point here is that a Turing Machine just does two things. It reads a number, and then based upon what it read, at the very end of the list it produces a number. Then, it slides over, and reads the next number, and that tells it what to write again. That's all a Turing Machine really is.

In fact, let's look at an example together, just to get our feet wet, with what this Turing Machine business is all about. It is not at all difficult, but of course, as we have seen so many times in this course, just because something is simple doesn't mean that somehow it's without content or without great value. In fact, this is really the beginning, and in fact, holds the structural, fundamental ideas for the modern computing that we do every day. Even though this is simple, then, it's quite profound.

Anyway, here we can see, for example, another little ticker tape, if you will, of numbers. It starts up "1, 2, 0, 5," and that's the sum computer program output based on some little program. Let's take a look at this little simple Turing Machine we have in front of us here, though, and here's the machine, or if you wish, here's the actual program. It says that if we read a 1 anywhere, what we are to put the end of our list are the numbers 2, 1, 2, so we're going to write those three numbers down. If we see the number 2, we're going to record the numbers 1, 3, 0; and if we see the number 3, we're going to record the number 0.

This is, in fact, a very simple Turing Machine, where, when we see 1 at the end of our list, we put the numbers 2, 1, 2. When we see a 2, we put 1, 3, 0. And when we see 3, we put a 0. That is an example of a Turing Machine or a program, and we can run it, so let's actually run it, and see what happens. We're going to start with 1, so let's pretend that the given input, in fact, is 1. We read 1, then, and what do we write? Well, when we see a 1, we write 2, 1, 2, so we write "2, 1, 2." Then, we slide over, and we see 2. When we see 2, we see that we write 1, 3, 0, so now, at the very end of our list, we write "1, 3, 0." Then, we slide over, and we see another 1, and when we see a 1, what do we write? We write 2, 1, 2, so at the end of our list, we write "2, 1, 2." Then, we slide over, and we read a 2. What do we do? We add on "1, 3, 0" at the end of our list. Again, we slide over to the next 1 we see, so we write "2, 1, 2" at the end of our list. We slide over again, we see a 3, and so we write a "0" at the end of our list, and then we slide over one more time, and look what we read. We read the number 0, so that now, we're reading 0. What does this computer program tell us to do when, in fact, we see a 0? Well, the answer is "nothing," because if we see a 1, we do something; if we see a 2, we do something; if we see a 3, we are told to do something; we don't know, however, what to do when we see a 0. That wasn't given, so what happens? The program stops, so in fact, that's the end of the program. The program terminates.

Now, in fact, this actually is an illustration of an extremely important theory in the notion of Turing Machines, which is determining, just given a program, that program will terminate or not. Will it stop? In fact, this is called the *halting problem*, because the question is: Does the computer halt? Does the program actually come to a stop? Now, actually, computer scientists call these things "halting." You and I would call this "crashing," because when things like this happen, the computer crashes, but in the computer scientists' lexicon, they refer to this as "halting." In fact, though, what's happening is that the computer program crashes. It stops because it doesn't know what to do. It read a 0, and we never told the computer what to do when, in fact, it sees a 0, so that's stopping. In fact, a very difficult question in computer science is if I give you a program to look at and to determine if it's going to stop in a finite number of steps or not. This is an example of a computer program, a Turing Machine, where, in fact, it does stop in a finite number of steps.

Okay, very good. Anyway, I hope you're warmed up as to what Turing Machines do. It's not very hard at all. They're very simple, of course. Let's take a look at another example, just for fun. This is another Turing Machine, so again, I've got to give you the rules. Let's take a look at it, and here are the rules. If you see a 0, write down "0, 1." If you see a 1, write down "3, 2." If you see a 2, write down "4, 2." If you see a 3, write down "3, 1," and if you see a 4, write down "4, 1."

Well, before we actually run this computer program starting with the input of 1, let's just see if we can answer the halting question. Will this computer program at some point actually halt? Well, I actually claim that we can answer this. It's not very hard, because look, we know what to do if we have any input of the form 0, 1, 2, 3, or 4. If we have any of those inputs, we know what to do.

Now, what are the outputs? What are the things that we produce? Well, if you look in the list of pairs of numbers that we output, it's 0, 1, 2, 3, or 4. Whatever we put out, then, we know that we will be able to read later. In fact, then, this is an example of a program, a Turing Machine, where, in fact, we know that it will never quit. It will never stop running, because every time it reads some old output, either 0, 1, 2, 3, or 4, we know that it can absorb it and produce something else further down the list, so this process will never stop. This is never going to stop.

Now, let's take a look and run the program, and see what happens. We will start with 1, and what do we do? We look on the computer program, we see a 1, and we are supposed to write "3, 2," so we write "3, 2." Now, we slide over, and we read a 3, so what do we do? We write "3, 1." Now, we slide over, and we see a 2; what do we write? We write "4, 2." We slide over, we see a 3, and we write "3, 1." We slide over, and we read a 1, so we write "3, 2." We slide over, and we see a 4, so we write "4, 1." We read a 2; we write "4, 2." We see a 3; we write "3, 1." We see a 1; we write "3, 2." We see a 3; we write "3, 1." We see a 2; we write "4, 2." I can do this really fast. We read a 4; we write "4, 1," and now, I am off the page here, so I'm going to stop, but this keeps going, and it's never going to stop. It goes on and on forever.

The output of this particular Turing Machine, then, is that long ticker tape of numbers. It's "1, 3, 2, 3, 1, 4, 2, 3, 1, 3, 2, 4, 1, 4, 2, 3, 1," and so on. Well, not particularly interesting, and maybe you're not really that impressed, and I admit that it doesn't look that interesting. Actually, though, I would argue, in fact, that those numbers, even though they are just from 0 to 4, are too complicated. They are just too complicated, so I'm going to propose that, in fact, we make those numbers simpler. Let's just look at those numbers, and just ask, "Are the numbers even, or are the numbers odd?" Let's just record that.

If a number is even, then, for example, 0, 2, or 4, let's just replace it by 0. Let's just say that that is going to be a 0, and if the number that we see in the output is odd, for example, a 1 or a 3, let's just record that as a 1.

Now, let's look at our output, and now make this conversion, to make it actually be a little bit simpler, and what do we see? Well, what we see here is "1, 1, 0, 1, 1, 0, 0, 1, 1, 1, 0, 0, 1, 0, 0, 1, 1, 1, 0, 1, 1, 0, 0, 0, 1," and it goes on. But now, let me put on this graphic of the paper-folding sequence, in terms of zeros and ones, and look at the amazing coincidence. They agree. In fact, what we see here is that the paper-folding sequence is really the output of a Turing Machine. So that in fact, these very complicated folds that we were looking at really come from the simplest kind of computer program you can have. Notice that in this context, we do not need have the previous answer to build the next answer. We just start right at the beginning, and we keep reading those digits off, and they tell us what to stick on at the very, very, very end. We keep reading and writing, and reading and writing. We can do that forever, and produce this infinitely long paper-folding sequence all from this very simple computer program, a computer program that really only has five lines in the program.

Look at the power of this Turing Machine idea. It allows us to, again, come back to this really chaotic mess, and tame it and make some sense. We also see, now for the first time in our experience together, that this chaotic mess doesn't have to be generated by the previous chaotic mess. In fact, we can just start right at the beginning, and start reading off those things, and they themselves are the instructions for what to write at the very, very end. As we traverse up-and-down, that movement, in fact, tells us what to do miles, and miles, and miles away at a horizon we're even yet to fathom, but by our actions here, we're setting the stage for our journey later, which is really a wonderful idea through this Turing Machine notion.

Well, there's a fascinating connection between the dawn of the computing, the fundamental ideas of computing, and folding a piece of paper, so that is pretty remarkable. In the last few minutes together, I want us now to take a look at this paper-folding sequence, but in its own right, just visually. Remember, this is an entire realm about visual things, paper-folding to fractals, and what I wanted to do was to return to these objects, and so there they all are. I will attend back on here so that you can admire them again. You can make them yourselves. This is just, of course, the various levels of paper-folding as the folds get more and more complicated. This is just the paper-folding with one fold, two folds, three folds, and so on. This is a bunch of folds. I think this might actually be five folds. Five folds there; you can enjoy that.

Now, what I'd like to do is—this, in fact was the flying bird that you might recall, actually, I don't want us to think about flying birds, but I want us to look visually at these things, visually, and instead of having them lay here like that, I'd like to have every crease now be at a right angle. Let's have every crease be at 90 degrees, so, for example, this bird, now, would actually become this right angle piece right here. It would look like that. That's the image that I want us to now think of when we think of the first fold.

Now, this fold, too, of course, can be made if all the angles are right, and let me actually show you how that would look. It would look like this. I will make it a little lower, so that you can see it a little bit better, there. In fact, therefore, if you enjoy origami, you can call this the origami gutter. Do you remember how gutters look, sort of like that? Or, an origami pot? Here's the handle of the pots, and you can put the water in here. You can make a lot of origami; in case you are origamically challenged, these are no problem. You have wonderful origami to show your friends.

For our purpose, though, what I want to do is just look at this, come and see it as a collection of right angles that come together, and if you look closely—and here, maybe I'll try one more just very quickly here; I think it is going to be tricky for me to actually show you; if I slide this one down, maybe it will be okay. There we go. You can take a look at that and see that, in fact, here's the image that we look at, all the angles being right, so now I'm actually trying to look at these objects, and if we look at this carefully, what we see is something interesting.

What we see is that we can return to the generating process that we saw sometime back, where what we do is take a particular sequence of paper folds, and spread them out, and then add the "over, under, over, under, over, under," and by adding the "over, under," we actually generate the next term in the sequence.

Now, what would we do in this case? Well, if we look at this—let's just look at this particular configuration, and try to add in the "over, under, over, under." If we were to add in a fold coming down this way and then going up, and then a fold going out and coming in, we would actually see this picture. In fact, let me see if I can actually show it to you by lifting it up, and trying to put in its place. If I hold it like this, what you see is the original "L," but here, I dipped below, and then here, I dipped over. Notice that if I am just traveling along the "L," I've gone to the right, and then when I come back, I go to the left, so that the move of going "up, down, up, down," has become right, and then left.

There is the alternating sequence, to produce from the "L" to this configuration. See? Let me just show you that one more time, so that you can really see that what we have here, in fact, is that I've gone to the right, and if I keep journeying back, I go now to the left, and I added that in, and that's the sort of up and down feature that we saw, which was precisely the next stage in producing the next sequence.

Now what I would like to do is notice that that process can exactly be continued. For example, here, if I take this right angle version that we have, and I do the same process, I go to the right, and then I go to the left, and then I go to the right, and then I go to the left, that weaving back—remember the weaving back we did before? If we do the weaving now, but we stay on here, we actually produce the next one. In fact, then, we can actually write down at the right angle version, pictures of these, exactly what they would be, by just taking the previous one, and kind of serpentining through them, going to the right and then to the left, and then to the right, and then to the left, including between each of the current folds, in which there are new folds, either folding this way, or folding that way, depending, in turn.

Well, this is wonderful, and you can sort of see reflections of this come, and in fact, if you were to do it for the next one, you would get an image. Let me just do it for you right now, live, and you can also kind of see it on the screen. There's an image of it on the screen, I suppose, but let me just make one here live for you, and make it at right angles. I am going to lift this up so that you can see it, and interesting thing that you notice is that if I make this perfectly at right angles, what you can see is that there is this square that is actually a complete square. Let me hold it really tight so that you can see it, right here. Do you see how that square actually is a square? Well, of course, it is just barely touching at that corner there, but it is a perfect square. All the angles are right angles, but it is just touching. Even though it looks like, in fact, it's a closed square, it is actually that one piece, in fact, that comes back onto itself like that, and so the image, as you can see on the screen, has, in fact, that little square feature, but it is still a piece of paper that has been folded.

Let's take a look at these images, and see that, in fact, this process can be continued. Therefore, if we start with that L-shaped piece, then, we can get to the pot by just merely serpentining back and forth, "right, left, right, left," and you see how we can build the pot. If we look at the next animation, we can start with the pot, and then go "right, left, right, left," alternating through the different corners of the pot to produce the next image, the image that is in the far bottom on the stand, on the easel. Then, we can repeat this process, and you can see the image, now, that we have, consists of that little square piece. Now, that square piece, of course, is just the pieces of paper touching perfectly, and they form that square, but of course, it's not just that little solid square there. If you were to move the paper a little bit, that square would open up just as it does here. That square would open up, and it would be apart. When you make everything perfect right angles, though, they just touch, and you have that meeting of the minds, or, in this case, of the vertices.

Well, we can certainly keep doing this, and so forth, and, in fact, this has actually has been done, and appears in literature. It is kind of fun to see literary reflections of some of the mathematics that we're looking at. If you look into Michael Creighton's book, *Jurassic Park*, which, of course, is a famous movie, look at the large print edition of the book, and you will see that in fact, the book is divided up into various iterations. They are rather like meta sections. There are chapters, but then there are these meta sections, as you know, *Jurassic Park* is the story of these scientists that now have these eggs, and so forth, and they have this farm where they are growing these dinosaurs, and all of a sudden, things go crazy, and the dinosaurs want to hurt the people, and people are really, really scared, and rightfully so.

Anyway, let's take a look at the actual book as it physically opens up, and let's look at the different iterations. For example, the first iteration—we see the page here—opens with this curve, and that curve is exactly this paper-folding curve. It is exactly this one, and you can see it; in fact, I am holding it up right here. You can see that it matches with the image, and then on the bottom, there is this quote by the mythical character Ian Malcolm, and he says the following. This is the quote from the book: "At the earliest drawings of the curve, few clues to the underlying mathematical structure will be seen."

That was Ian Malcolm's quote from the beginning, and so far, everything is fine with Jurassic Park. Then, in the next iteration, all we are going to do, if you look at that image here, is to just do the next paper-folding sequence, so you go "over, under, over, under," in the previous one, so he therefore uses paper-folding in his book to illustrate how things get complicated.

Now, in the next iteration, he says: "With subsequent drawings of the fractal curve, sudden changes may appear." Now, if we look at the third iteration, it's just another iteration of the paper-folding curve.

Well, what happens here? Well, look what happens. We see lots of little squares, but we understand that those aren't really squares, but the paper coming together, fitting together in a very complicated way.

"Details emerge more clearly as the curve is redrawn," Ian says. Now, the dinosaurs are getting out of control, and it's getting sort of scary, and look at the next image, the fourth iteration. It is looking more complicated and more square-like, but also, maybe a little bit more familiar. "Inevitably, underlying instabilities begin to appear," the character says, and you can see by the fifth iteration, that the pattern begins to look very familiar.

"Flaws in the system will now become severe," the character says, and if we move to the sixth iteration, there's no doubt about it. We have found the Dragon Curve. By taking the paper folds, and unfolding them so that every angle is a right angle, suddenly we find that beautiful, infinitely complicated fractal Dragon Curve from the intermission at the end of the last lecture, finally tying up that last loose end, and seeing fractals. Ian says: "System recovery may prove impossible." Finally, in the seventh and last iteration of the book, you see a beautiful rendition of the Dragon Curve, and in fact, the last quote from this opening is: "Increasingly, the mathematics will demand the courage to face its implications," and what a wonderful quote for this mathematician character in the book to say, because that is exactly the reality, that mathematics brings us to a point where suddenly, we have to face those implications.

Seeing reflections of the Dragon Curve inside of and from just folding a sheet of paper is amazing. If you think back to when we first saw the Dragon Curve out of context, what we saw, in fact, was a curve that seemed impossible to explain. It was too complicated to explain, and yet, what we discover now is that it is as simple as folding a sheet of paper.

Well, I think this journey was just wonderful, because it offers us a powerful illustration of how mathematical patterns and structure illuminate a process, you know, so that our quest for understanding structure has taken us on a wild journey, right? From paper-folding to pattern recognition, to the naive arithmetic, to the evens and odds, to the algebra of generating functions with those fancy equations, to Turing Machines, and to Dragon Curves. We see that mathematical realms truly have a life of their own, and in order to understand and to uncover many of those unexpected worlds, we have to think literally outside of the box. Those worlds usually offer us incredible intellectual vistas to enjoy, and once again, we see the power of looking at simple things deeply, and searching for a pattern. In the next lecture, Mike will pick up the idea of fractals and we'll see fractals appear both in the abstract and in our everyday world.

Lecture Seventeen
Fractals—Infinitely Complex Creations

Michael Starbird, Ph.D.

Scope:

Some artists are renowned for their attention to detail, producing pictures that are incredibly intricate. But only within the realm of mathematics can one create images that are literally infinitely detailed. What does it even mean to speak of an image that is *infinitely* detailed? Could such an image be drawn with the finest ink? No; instead, we describe such images so precisely that for any point on our canvas, we can determine whether that point is in the design or not. These images can be drawn to any degree of detail that our finest printers can maintain. But the totality of their intricacy is only fully present in our mind's eye. Every power of magnification reveals yet further detail. These infinitely detailed images arise from repeating a simple process infinitely often and reasoning about the result. The beautiful images so created illustrate ideas of self-similarity and symmetry.

Outline

I. One of the best strategies for coming up with new ideas is to extrapolate a familiar property in the environment to a more general setting.

    A. In this lecture, we discuss some observations we can make starting with broccoli.
        1. Note that one small stalk of broccoli looks like the whole bunch.
        2. We can repeat this discovery about broccoli; that is, an even smaller stalk again looks like the whole bunch.
        3. Broccoli has the property of *self-similarity*.

    B. We can also see this property in maps.
        1. The coast of England, for example, looks jagged on a map.
        2. A more detailed map of a part of the coast looks similarly jagged.
        3. In other words, a small part of the map looks similar to the whole.

- C. As we explore the property of self-similarity, we will discover the world of fractals.
  1. An example of a fractal that appears in the real world can be seen on a Quaker Oats box.
  2. The box shows a picture of a pilgrim holding a box of Quaker Oats, which has a picture of the same pilgrim holding a box of Quaker Oats, and so on.
  3. Conceptually, that repetition would occur forever.
- D. If we zoom in on any part of a fractal, such as a Julia set, we see that any part of it contains as much detail as the original did. Fractals are literally infinitely detailed.
- E. What is at the root of these beautiful objects of infinite complexity, and how do we create them?

II. We begin our exploration of fractals by imagining the creation of a carpet with a design based on a square.
- A. Picture a square carpet with a purple background and an orange square in the middle. We can divide the carpet into a three-by-three grid, which shows that the orange square is, in essence, surrounded by eight squares of the same size.
- B. Further suppose that we change the eight surrounding squares to match the overall design of the whole carpet. That is, each smaller square will be divided into a three-by-three grid and the center square will be colored orange.
- C. Does the whole carpet now look exactly alike? No; even as we repeat the process of dividing squares, we find that we are always one stage behind. Each surrounding square doesn't look quite like the whole carpet.
- D. How, then, could we make each square of the carpet look exactly like the whole? We would have to repeat the process of dividing squares and coloring the center square forever.

III. Let's try this method of replacement with a straight line segment.
- A. Let's replace the straight line segment with four segments, each one-third as long as the original segment. The result is a jagged line.
- B. Now let's repeat that replacement process. We replace each line segment with four segments, each one-third as long as the segments in the second iteration. Now we have 16 jagged parts making up the line segment.

- C. If we continue this process, we will start to see that the resulting jagged segment looks very similar to itself.
- D. The object that we have created is called the *Koch curve*. The whole is exactly identical to any one-third of the curve.
- E. We can assemble six of these curves in a circular pattern to make a Koch snowflake.

IV. Another method of creating examples of fractals is to make a collage.
- A. We can put together a collage by making several reduced copies of a picture and arranging those copies artistically on a page.
- B. We then take the resulting collage and repeat the whole process.
  1. That is, we make the identical number of reduced copies of our collage and arrange those reduced images in exactly the same way as we did the original pictures.
  2. In this fashion, we create a collage of collages.
- C. Now, we repeat the whole process, making a collage of collages of collages, and so on, forever.
- D. Let's try this collage-making process with a triangle instead of a picture.
  1. We start with a triangle, make three half-size copies of it, and arrange the copies in a triangular pattern with a blank space in the middle.
  2. We then repeat this process infinitely. To understand this concept of infinite repetition, note that there are some points in the image that are in every iteration of the triangle. The result is a famous fractal called the *Sierpinski triangle*.
  3. This image is infinitely detailed, and each one-third of the image is identical to the whole.
- E. We can create another fractal, the *Sierpinski carpet*, using the collage process.
  1. This time, we start with a square; make eight reduced copies of it, each one-third the size of the original; arrange the copies around the original square; and repeat forever.
  2. As before, we get a fractal in which each of the eight surrounding squares looks identical to the whole.

V. This collage method can be used to create fractals in higher dimensions, as well. In this case, we make a collage of sculptures.
   A. We begin with a cube and reduce it to one-third the original size in each dimension.
   B. We make 20 copies of this reduction and arrange them in such a way that we could drill holes and see through the structure in any dimension.
   C. We then repeat the process, making 20 reduced copies of the whole and arranging them as before.
   D. The result of repeating this process infinitely many times is the *Menger sponge*. It resembles a sponge because it has many holes that allow us to see through the structure.

VI. In creating fractals, the original image is immaterial. Only the size of the reduction and the arrangement of the copies are important.
   A. For example, we can make the Sierpinski triangle with a square. In this case, we make half-size reductions and arrange them in a triangular pattern. If we continue this process, the result starts to look more triangular.
   B. We could also reduce and arrange circles or any other object to form a Sierpinski triangle.
   C. Indeed, we could begin with a Sierpinski triangle and reduce and arrange copies of it to form a Sierpinski triangle.

Suggested Reading:

Barnsley, Michael. *Fractals Everywhere,* Academic Press, San Diego, 2000.

Burger, Edward B. and Michael Starbird, *The Heart of Mathematics: An invitation to effective thinking*, Key College Publishing, Sections 6.1, "Images: Viewing a gallery of fractals" and 6.3, "The Infinitely Detailed Beauty of Fractals: How to create works of infinite intricacy through repeated processes."

Questions to Consider:

1. Koch's curve is created by starting with a straight segment and replacing it by four segments, each one-third as long as the original segment; so, at the second stage, it has three bends. At the next stage, each segment is replaced by four, and so on. How many bends does this curve have at the third stage? The fourth stage? The nth stage?

2. Can you position three mirrors in such a way that, in theory, you could see infinitely many copies of all three mirrors? Either describe the positions of the mirrors or describe why it is not possible.

Lecture Seventeen—Transcript
Fractals—Infinitely Complex Creations

One of the best strategies for coming up with new ideas is to look around the environment and see some property that is familiar, and then extrapolate it to a more general setting. By doing that we can develop whole new ideas and discover all sorts of amazing consequences.

In this lecture, we're going to be talking about an observation that we can make, starting with a very simple, everyday thing, which some of us find less tasteful than others, but all of us find healthful, and that is broccoli. We will start with this very familiar object, broccoli. Now, this is a bunch of broccoli. When we were younger, and our mothers were telling us to eat our broccoli, of course, we wanted to advertise that we had eaten a great deal of broccoli, because we didn't really like it, but we wanted to eat a lot of it.

Broccoli, though, was good for this purpose, in that there is a strategy that you can use to eat just a little broccoli, and make it seem as though you've eaten a lot. The way you would do it is this: You would notice that this big amount of broccoli—that is a huge amount of broccoli, and, of course, this is much more than anybody would want to eat; the good thing about broccoli, though, is that if you look at just a part of the broccoli, for instance, just this much here, it really looks very similar to the whole bunch. Particularly if you look at it very close, it looks like the whole bunch. It has the property that a small piece of it looks the same as the whole bunch. In fact, we can iterate this process, that is, we can do the same thing again, and look at an even smaller piece of broccoli, like this right here, you see, and it looks very much like the whole bunch. In fact, if you took a close up photograph of this little piece of broccoli, it would be very difficult to distinguish from this big bunch of broccoli, and in fact, we can go even further.

Look at this little tiny piece of broccoli. Now, this little tiny piece of broccoli really does look very much similar to the entire piece of broccoli. It is *self-similar*, meaning that the whole piece has little parts that look similar to the whole thing. Now, what you could do is just eat this little piece of broccoli, and then take a photograph of it, blow it up, send it to your mother, and she would think that you are healthful. I'll do that.

This is an inspiration for thinking about the concept of self-similarity. There are other places in life where we see self-similarity. One is in looking at maps. If you take a map of England, and you look at the shoreline of the map, you'll see that it has a very jagged kind of look to it. If you then zoom in on it, though, and look at just a part of the shoreline of England, you'll see that it also has that same sort of jagged character. Once again, then, what we are observing is a similar feature in broccoli and in the maps of England, in the sense that a small part of it looks similar to the whole.

Now, what we're going to do is to practice a strategy of taking an observation like taking something that looks the same in a small part of it as it does in the whole, and exploring where it can lead us. In this case, it is going to lead us to fractals.

Now, fractals are beautiful objects, and they have this concept of self-similarity. What we're going to do is talk about how these fractals were constructed by using the process of repeating an easy process over and over again. There are some examples of fractals that you actually see in real life that have this self-similar property that occur on packaging of various things. If you remember back in the old days, there used to be Quaker Oats boxes where there was an image of an old gentleman from the Revolutionary Era days holding a box that was a picture of the entire Quaker Oats box. Well, of course, his picture was on that box holding a Quaker Oats box, and so, once again, it had his picture on it with an image that continued.

Now, of course, it really didn't continue forever, because there's a limit to how fine ink can be, but in fact, there's a concept here that that repetition would, in fact, conceptually occur forever. And so you would have infinitely many people holding Quaker Oats boxes inside Quaker Oats boxes, inside Quaker Oats boxes, that would actually go on forever, so that the image would be infinite.

We can see some really complicated fractals that you see often on calendars, and so on, that have this self-similarity to them, and you'll see here, in this fractal—this is called a Julia set actually. It is an intricate fractal, and we can zoom in on part of that fractal, and you'll see that fractal contains as much detail as that entire first fractal did. Fractals have the property that they are literally infinitely detailed, and to me that concept in itself, the idea that something can be infinitely detailed, not just extremely detailed, but literally infinitely detailed, is an amazing idea, that an image can be constructed in the mind. It no longer is an image that is drawn, but is an image in the mind, so that when we take a fractal, what it means is that we know which points are in the fractal and which are not, even though you couldn't possibly draw them.

We can therefore magnify this fractal billions of times, and see what it would look like if we were able to see a billion times its magnification, and in some of these fractal images that you are looking at now, you can actually zoom in many, many times, and you'll see that it continues to have infinite detail even at these great magnifications.

The strategy then of the lecture is to tell you what it is that is at the root of this infinite complexity and how we create these beautiful objects of infinite complexity. We will begin with the story of a carpet maker. This gentleman was interested in making some beautiful carpets, and he had an idea that he was particularly attracted to, the very attractive pattern of a square. A square is a beautiful shape, everybody likes a square, and so he wanted to make a pattern for a carpet that accentuated the square. In order to do that, he started off with a square; but of course, if you just make a square of all the same color, it's a little boring. He wanted to accentuate the property of the squareness, and so, in the middle of the square, we'll put a little, different colored square—in this case, a orange-colored square in the middle of this purple background.

Well, this accentuates the square, and it creates not only an emphasized square in the middle, but there are also eight squares around the outside that are the same size. You see, he took the big square, divided it into a 3-by-3 array like a tic-tac-toe board, and then changed the color of the inner one. There are then eight squares surrounding this center square.

Well, suppose you wanted to make a carpet so that each of those eight surrounding squares had as much attraction to them as the whole square did. You see, each of those eight surrounding squares had the same defect that he had found in the original square, namely, that it was boring; it was all just one color. In order to correct that defect, then, we could go about cracking it in the same manner that was originally done to correct the defect of this original, big purple square. Namely, to take in each of those eight surrounding squares, the center, divide it into a 3-by-3 grid, and take the center third of that, and make it the different orange color. This creates a square that has eight surrounding squares that each have one orange square in the middle, and then a big orange square in the center.

Now, we might want to just rest, and say, "Well, now, I've done it. I now have a square where each of the eight surrounding squares are the same as the first," but if we look at that, we see that, in fact, it's not the same anymore. It's not the same, because each of the eight surrounding squares has just one orange square in it, whereas now, the whole picture has a big surrounding square in the middle, and then eight smaller orange squares surrounding it.

Well, how can we correct that defect? The way to correct the defect is to just go ahead and fix it in the same way that we fixed it before. Namely, for each of those surrounding squares, we want to make that little square be the same as the big one now is, and in order to do that, we take the whole picture and put it in that one little square. Well, now, that creates a picture that looks like we made progress in the sense that each of the eight squares surrounding the middle one now has a big square in the middle, and eight squares surrounding it. Now, we can rest. Now, we're done.

Well, not quite. We're always one off, because now the whole square doesn't look quite the same as one of those eight squares around. It is one iteration off. In order to make each of those eight surrounding squares the same, we have to correct each of the little squares and put a bunch of additional orange dots. What, then, is going to be our strategy for correcting this and really making each of the eight squares exactly the same as the whole thing? The answer is that we just do it forever. It is not enough to do it one time, ten times, or 100 times. We do it literally forever, and after we've done it forever, we'll have an object with the property that each of the eight squares surrounding the center one is absolutely identical to the entire big square, except at a smaller scale.

Now, this is a wonderful example, and it gives us a strategy, and once we have a method for constructing something beautiful, then, we can apply it in all sorts of different settings. Let's try it again. Let's just start with a straight line segment. We start with a straight line segment, and since it's a little bit boring, we can replace that straight line segment by four segments, each a third as long as the original segment. We start with one segment, and we cut it into three pieces, but instead of having just three of them, we have a third, and then, we have to make little bit of a jagged part to come up, because each of those is a third as long as the original segment.

This then gives us this first iteration of a bumpy line segment. Let's do the same replacement process again. That is, we take each of the line segments, replace each of those segments by four segments, each a third as long as the original one. When we do that, we get this much more bumpy sequence. Now we have 4×4, or 16 little segments, and you can see how bumpy it is, but it has sort of a nice star-like quality to it. Once we get in the habit of repeating, let's just go ahead and repeat. That is, each of the small segments we now have, we replace by four segments, each one-third as long as the original, and we continue this process again, and again, and again, and pretty soon, the resolution of the TV screen, or of any inkjet printer, or of any kind of actual drawing device, is exceeded, and we will start to see that it looks very similar to itself.

In fact, one thing that we notice is that the whole curve that we get, the whole object that we have now created, which incidentally is called the *Koch curve*, is exactly identical to one-third of the curve. So that, if you look at just the first third of the curve, that final picture is exactly the same as the whole curve, except one-third as big, and in fact, there are four of them. That third, the third that sticks up, here, the one that sticks down, and the one at the end—there are four of them that are each exactly the same as the whole curve except one-third as big. In fact, each of those, of course, is exactly the same, so that if we looked at one-third of it, that thing, the very small one, would look exactly the same as the entire Koch curve. This, then, is an example of an object that is self-similar exactly at infinitely smaller scales, that is, at infinitely many different smaller scales; one-third as big, one-ninth as big, one twenty-seventh as big, and so on.

I wanted to say that this curve also—since we started out with something of unit length, and then we made something that was four-thirds as long, since we had four segments each a third as long, then in the next one, each segment was four-thirds as long, so that the total length at the next iteration was four-thirds squared, $\left(\dfrac{4}{3}\right)^2$, and then at the next iteration, four-thirds to the third power, $\left(\dfrac{4}{3}\right)^3$. You may remember the exponential growth that we encountered in Lecture Two, that that curve goes to infinity in length, so that the final curve actually has infinite length. We can actually put this curve together into a snowflake by assembling six of them around to make a Koch snowflake.

Well, I wanted to demonstrate another method of creating these beautiful fractal images, and that is, instead of doing a replacement process as we did there, let's do something that is more artistically valid, and that is to start with a process that we all use when we're making photo albums. Now, the way to make a really attractive photo album is to make a collage of pictures, because you take pictures that are very attractive, and then by arranging them artistically on a page, you make something that is even more attractive.

What we are going to do, then, is to do that process of making collages, and we will see that in doing so, we can make a fractal. Let's start with making a collage in the following way. First, we take a picture that we find very attractive. Well, okay, here is one. Isn't that attractive? Well, maybe not. In any case, this is the picture we will use. So we will start out with this picture. Suppose we decided that this was so attractive that we decided to make a collage of it, and since this picture is the one that we really like, we would make the picture collage using only this photograph. But in order to fit it on the page, we would have to reduce it in size, and take several copies and assemble them, so that's what we'll do.

We'll take this photograph, and reduce the size, and make three copies. Here we go; this is the big photograph, and then here are three half-size copies. It is half in this direction, and half in this direction, up and down. So, we have these three things, and then, we make a collage out of it. We can make a collage by arranging them attractively in this manner. Isn't that an attractive collage? Here we have it all glued down. This is a collage of three pictures that are half the size of the original one.

Well, remember what we do when we have something that's attractive. We make a collage. What we do as we take this picture, the entire picture, which is a collage itself, and reduce it by 50 percent, make three copies, and arrange them into a collage. Here we have three copies of this picture, reduced halfway; there they are, and we can arrange them exactly as we did before, because this sort of triangular arrangement is so attractive. That's the arrangement we are choosing to use to make this collage. So let's go ahead do that, and when we do that, we assemble them to create this beautiful collage of collages. You see?

Isn't that attractive? Well, what do we do when we see something that is attractive? We make a collage out of it, and by the way, one of the features of fractal development and maybe mathematics in general is that once you have an idea, you just continue it over and over again, because often, it leads to neat ideas, particularly if you go to infinity, which is something we can do in mathematics, and maybe in real life, we can't quite get that far.

We take this picture; we make three copies of it, half size, you see? We arrange them in this nice collage shape and glue them down, and we get this beautiful collage of collages of collages.

Now, of course, what do we do next? The same thing again, the same thing again, the same thing again, forever. Forever.

By the way, when we get to this stage, you might notice that the pictures of this wonderful picture here are really rather small. You can't actually see the details of each individual picture very much anymore, and we'll see that, in fact, what picture you start with has very little to do with what it is you end up with. We'll think about that as we go.

Let's do an example that may be a little bit more attractive if you are a fan of triangles. So let's just start with a triangle, and do this same collage-making process. That is, we start with a triangle, make three copies of it that are identical in shape but half the size, and arrange them exactly as we did with these photographs: One in the middle at the top, and two on the bottom. And there, we get this triangle with a blank spot in the center. Now, we're going to repeat the process, because that's the strategy of the collage process. Once we have an idea, we repeat it.

We made a copy of that, three copies of half size, and arranged them in the same places. Then, we made three copies of that, half the size, arranged them in three places, and we iterate. That is to say, we just continue over, and over, and over again. When we have finished—by the way, to say that we have "finished" means that we literally do it infinitely many times, and what I mean by that is that there are some points on the paper that are in every single iteration of that triangle. They exist in the original triangle. They exist at the second stage, they exist at the third stage, they exist at the fourth stage, and they exist forever, and those are the points that are the ones we're talking about that persist to infinity, and that is the object that we get at infinity.

That object is called *Sierpinski's triangle*. It is infinitely detailed, and if you look at one-third of the triangle, of Sierpinski's triangle, that of the bottom left corner, the bottom right corner, or the top corner, each of those is identical, absolutely identical, to the whole picture. Think about why it has to be identical. It has to be identical, because if you look at that top third, how was it constructed? Well, it was constructed exactly the same, in exactly the same manner as the whole triangle was, except that it was one step behind, but since we take infinitely many steps, it also has endured infinitely many steps of this collage process, and therefore, it is exactly identical to the entire original triangle, the Sierpinski triangle. This, then, is a collage method for making fractals, so I like to think that what I am providing here is a collage education.

Another fractal that we could create using this method is going back to *Sierpinski's carpet*. Sierpinski's carpet is from the same gentleman, Sierpinski, who made this other fractal that we saw before, that started with the square. Once again, we could create it not by the replacement process that we did before, but by this iterative method of creating, the collage method of creating fractals. Because if you start with a square—then you say, "Oh, isn't that an attractive square?"—then we reduce it, make eight copies that are a third the size each, and place them around the original square. Then, that gives us the next iteration of the collage-making process for the Sierpinski carpet.

Then, we take that picture, take a photograph of it, reduce it to one-third the size, make eight copies, arranged them neatly around the center square as before, and that creates the next iteration of Sierpinski's carpet. Then, we repeat the process again and again and again, and you can see that that creates the Sierpinski's carpet that we saw before. It also tells us why it is that each of the eight squares around the center one are absolutely identical to the final total

Sierpinski carpet except at one-third the size. When I say "one-third the size," what I mean is a third in both directions, so, of course, a ninth the area of the whole Sierpinski carpet.

There's no reason why we have to restrict ourselves in this process to images that are on the plane. We could be sculptors instead of artists, and create fractals that are in three dimensions, or even four dimensions. Let's restrict ourselves to three dimensions for now, and create a fractal that starts with a cube. We will start with a solid cube then. This solid cube can then be—the collage analogy is to take the solid cube and make many reduced copies of it, and then arrange them in some artistic way, so that it is sort of like a sculptural collage.

Let's try it in the following way. Let's take a cube, and reduce it to a third the size in each dimension, so that, in other words, we have a cube that's one-third as long, one-third as tall, and one-third as thick, and make 20 copies of them. If we take 20 copies of that cube, and we arrange them around in a way so that you have a hole going through from the top to the bottom, from the front to the back, and from the side to the side, you have, then, a cube that you can look through in any of the dimensions, and if you count carefully, you'll see that there are exactly 20 cubes that you have there, and you have removed seven of the 27 smaller cubes that would fill up an entire cube.

Once you have decided that strategy of fractal construction, the die is cast. If you just now iterate the process—well, what does iterating the process mean? It means taking this object, which we now find attractive, and we want to make a sculptural collage of it, so we we'll do is take it, reduce it to a third of the size in each dimension, and make 20 copies, which we arrange in exactly the same 20 positions that these were, and this creates the next iteration of this three-dimensional fractal image. I say "three-dimensional," but what I really mean is that it is sitting in three-dimensional space.

Notice that it now has many holes through it, the bigger holes go from side to side throughout the whole thing, and then, there are smaller holes that are created by having taken the big picture with the holes, reducing it to a third the size, and then putting the copies in those 20 positions. You can see that you have something with yet more holes going through it. If we iterate the process again, we would take this whole, holey image, holey solid, reduce it to a third the size, put 20 copies of it together, to get yet another iteration. If we do this infinitely often, we get an object whose name is the *Menger sponge*. You can see that it has sort of a spongelike quality because of all the holes in it.

What I would like to do now is to think about the fact that there are some possibilities for creating beautiful art that requires no artistic talent. This is very important for some of us to acknowledge, and what I mean by this, in the sense of fractals, is that it turns out that in creating fractals, it does not matter what the original image is that you start with. It doesn't matter. It only matters how much you reduce the fractal, and where you arrange the reduced copies in making the next iteration of the fractal.

Let's see this. For example, in Sierpinski's triangle, we're going to start with something that doesn't seem triangular at all, namely, a square. What we're going to do, though, is to follow the exact same rule that we followed when we made Sierpinski's triangle. Namely, we took the square, reduced it to exactly half the size, and arranged them in exactly the positions that are specified, namely, the center top, bottom right, and bottom left, and by arranging those squares in exactly those positions, we get an iteration.

Now, that looks a little bit triangular, because it has sort of a square top and a bigger base, but it doesn't really have the sense of the Sierpinski triangle yet. Look what happens when we iterate the process. If we take three copies of that fractal, that image I mean, we reduce it to half the size, and arrange them in exactly those same three positions—now, we're getting something that looks more triangular.

Now, we will reduce it again. We take half the size, put it bottom right, bottom left, and at the top, and we're getting something that looks more fractal-like. Then we can do it again, and we can see that pretty soon, we're getting an object that is indistinguishable from the fractal that we obtained by starting with a triangle.

In fact, it doesn't matter what we start with. Let's give some other examples. Let's see some other examples. We could start with a circle, and just arrange them by the same pattern. We have the pattern of the circle, and we have three circles, one on the top, one at the bottom, and one on the side, and then nine circles. By taking three copies of those and iterating the process, we get the same

Sierpinski triangle in the limit. Now, we can do the same thing starting with just a squiggle, and we can do the same thing starting with any object at all.

Look what happens, by the way, if we start with the Sierpinski triangle. If we actually start with the Sierpinski triangle as the initial image, what happens if we reduce it by half, and we arrange it in the three places? We get precisely the Sierpinski triangle back again, so in the case of the Sierpinski triangle, the iterated process just leads us back to where we started, but what we have seen is that it doesn't matter what the initial image is. We are always going to get the same result after we do infinitely many steps.

This, then, is an interesting insight that shows that even if we start with a picture that maybe not everybody agreed was extremely attractive, the final product will be the attractive Sierpinski triangle.

Lecture Eighteen
Fractal Frauds of Nature

Michael Starbird, Ph.D.

Scope:

The fractal images created by the collage-making process can resemble reality to an uncanny degree. We will see Barnsley's fern, which looks as though it had just been plucked from the garden, but in fact, it never existed and is the result of a simple mathematical process. Repeating the collage-making instructions creates intricate images, but another way to generate such images is by random luck. We demonstrate by playing the Chaos Game.

We start with an example that shows how a random process can lead to a definite picture. Similar random processes give an alternative way to construct infinitely detailed images that cannot be easily distinguished from photos of the real world. In fact, there are theories that the stock market and even one's heart rate exhibit the type of behavior illustrated by the fractal construction process with randomness thrown in. Thus, we see that chance, together with some simple rules, leads us to an infinitely intricate world of fractals, which quite possibly, overlaps with our own physical real world.

Outline

I. Remember that fractals are infinitely detailed and complex images that exhibit the property of self-similarity.

  A. We saw in the last lecture that we can create fractal images using the collage process, an infinite repetition of reducing and arranging images.

  B. Last time, we used the reductions of the same size to create collages, but we can also make collages using images that are reduced to different sizes.

  C. As we see, even if we start with an unattractive picture, the repeated process of collage-making yields a beautiful image. The original picture begins to disappear.

  D. In this case, the process of collage-making begins to resemble a fern. The image is called *Barnsley's fern*.

E. This image of a fern is so natural and lifelike that we wonder if nature uses processes that are similar to our process of iteration.

F. We return to the landscape pictures that we saw at the beginning of the lecture. These, too, are fractals—the results of a simple mathematical process repeated.

G. In fact, any picture can be approximated by a fractal; this method is sometimes used in data transmission because the instructions for generating an image can be written in just a few simple formulas.

II. Another method of creating fractals involves randomness. We'll learn how to create a specific picture by a random process.

A. Recall the Sierpinski triangle, in which each of the three parts of the image was identical to the whole image. We will follow a random process to arrive at this specific image.

1. Start with three points that make the vertices of a triangle and label them 1, 2, and 3.
2. Roll a die and move halfway toward the vertex with the same number, 1, 2, or 3. Mark that point.
3. If we repeat this process many times or program a computer to repeat it, we see that what starts as a fuzzy image begins to sharpen and resemble a Sierpinski triangle.

B. Why is it that this particular image is the result of this random process?

1. The first idea to understand is that if we start at, for example, the bottom left-hand vertex, that is a point in the Sierpinski triangle.
2. As we move halfway to vertex 2, imagine that we take all the points in the whole Sierpinski triangle and move them halfway toward the apex of the triangle. Remember that the top one-third of the triangle is identical to the whole but smaller. If we move up to the halfway point, we are constructing the whole Sierpinski triangle but at half scale. Again, we are still on the Sierpinski triangle.
3. If we roll another 2 on the die, we move the top half of the triangle into the very apex, the top one-fourth. Our point, which is three-quarters of the way toward vertex 2, is still in the Sierpinski triangle.

4. So far, we can conclude that as we roll the die and mark the points, we are always staying in the Sierpinski triangle. We never land in the blank space in the middle, for example, because that is not halfway toward vertex 1, 2, or 3, starting at any point in the triangle.

C. Is it possible that we could throw the dice infinitely many times and never reach a certain point on the Sierpinski triangle, for example, the bottom one-ninth of the triangle near vertex 3?
1. We will always reach that point if we roll the die infinitely many times.
2. We know for a fact that in rolling the die, we will eventually roll two 3s.
3. No matter where we are in the Sierpinski triangle, when we roll the first 3, we move toward the bottom right-hand third of the triangle. If we roll another 3, we move to the bottom one-ninth of the triangle.
4. We can refine this process even further and show that we will always land in any designated 1/27 of the Sierpinski triangle, because at some point, we will roll the appropriate sequence of three numbers. We demonstrate this phenomenon by selecting a 1/27 of the Sierpinski triangle and finding that rolling the sequence 2, 3, 1 will bring us into that small piece.

III. The rest of this lecture shows a strategy for creating ideas that is potent and applicable outside of mathematics.
A. If we find a situation that is beyond our understanding, a good way to approach it is to look at related simple things that we understand well, identify patterns in those simple things, and apply those patterns to the more complicated situation.
B. We can illustrate this idea using the Sierpinski triangle. We might note that the Sierpinski triangle does not seem to embody the familiar concept of dimension. We'll apply the strategy of looking for patterns to define a new concept of dimension for the infinitely complex Sierpinski triangle.

C. Think of a square, a familiar two-dimensional object. If we want to construct a square that is twice as big in both directions as a smaller square, we would need four copies of our original square. If we want to make a square that is three times as big, we would need nine copies of the original square.

D. We might also think about a cube. If we want to double the size of a cube in every dimension, we would need eight cubes ($2 \times 2 \times 2$).

E. What is the pattern? The number of units needed is equal to the number of times larger we want the original object to be, raised to the power of the dimension. Thus, to make the square two times larger, the formula is $2^2 = 4$. To make the cube two times larger, the formula is $2^3 = 8$.

F. We can also verify this pattern with a line segment. If we want to make a segment that is twice as long as an original, the formula is $2^1 = 2$.

G. How can we apply this pattern to the Sierpinski triangle?
1. How many copies of the Sierpinski triangle are needed to make a Sierpinski triangle that is twice as big as our original? Counting, we see that we would need three copies.
2. If we follow the earlier pattern, we get the equation $2^d = 3$; 2^d is the number of times larger we want the original Sierpinski triangle to be, raised to the dimension power, and 3 is the number of copies of the Sierpinski triangle needed.
3. If we solve the equation, the result is $d = 1.58496...$, a number that is between 1 and 2.
4. We now know a feature of a conceptual object that we derived from looking at familiar objects and applying patterns found there to the unfamiliar realm.

Suggested Reading:

Barnsley, Michael. *Fractals Everywhere,* Academic Press, San Diego, 2000.

Burger, Edward B. and Michael Starbird, *The Heart of Mathematics: An invitation to effective thinking*, Key College Publishing, Sections 6.1, "Images: Viewing a gallery of fractals"; 6.3, "The Infinitely Detailed Beauty of Fractals: How to create works of infinite intricacy through repeated processes"; and 6.6, "Between Dimension: Can the dimensions of fractals fall through the cracks?"

Questions to Consider:
1. Guess collage-making instructions that would result in a fractal that would look like a tree.
2. Compute the fractal dimension of the Sierpinski carpet and the Menger sponge.

Lecture Eighteen—Transcript
Fractal Frauds of Nature

As you know, I'm a mathematician, and sometimes mathematicians get a bad reputation of taking the life out of things. And instead of being attentive to beauty and the natural world, we're always trying to analyze them, break them down into pieces, and sometimes, people think of math as cold.

Now, I don't think of math that way at all, and to demonstrate my sensitivity to nature, I want to show you some pictures here that are just gorgeous vistas. You see this beautiful picture of a moonscape? Isn't that gorgeous, and look at these wonderful mountains, with the snow on them. These are just terrific images. Aren't these absolutely gorgeous?

Let's just take a moment to appreciate the beauty of nature. Well, okay, enough of the appreciation of the beauty of nature. Now, let's get back to fractals. Remember from last lecture that fractals are objects that are sort of infinitely detailed, infinitely complex kinds of images that can have a self-similarity property to them, that one part of it can look exactly like the whole thing. And we saw last time how we can create fractals by a method that we called the collage method of creating fractals, which was that we took a picture, and made a collage by taking reduced size copies of it and arranging them on page; and then taking reduced size copies of that image, and arranging them on a page in the same order; and then repeating the process forever, because, after all, we're now all mathematicians. We take things to the extreme in mathematics, and so, we create them forever, and at the end, we get these interesting fractal images.

Let's go ahead and do that with this example now. Let's take this image, which may not be too attractive, but we have this one image, and this time, we're going to make a slight variation on what we did in the last lecture. In the last lecture, we always took the same reduction of each copy that we used in making our collage, but sometimes for collages, we want to have one picture bigger and some smaller, and we will do it in this case. In fact, we will take this one picture, and we will make four reduced copies of it. The first copy we will make it about 85 percent as big as the original one, and we will place it at this kind of jaunty angle. Then, we will make two copies that are about a third the size of the original image, and we will put one on the right, one on the left, and put one kind of squashed together in the middle at the bottom. Now we have a new collage. So we will take that collage and make four copies of it. We will take one copy that is 85 percent of the original collage and put it at this jaunty angle, and then we'll make two other copies that are a third as big, put one on one side and one on the other—exactly the same places we arranged the original four pictures—and then we will take the fourth copy and squish it at the bottom. Now we'll take this process and do it over, and over, and over again.

Let's just see what happens as we repeat this collage process. We started with the first one; we got this next image with four reduced copies. Then we reduced them, got four copies to make a collage of collages, and then reduced it again, took four copies, made a collage of collages. By the way, you can see the collections as you go from image to image. You can see that they persist. That is, you can see the image of having a picture with pictures come to the side at this one, and over the side.

Now, let's take it again, and look at the next image, and look at the next image, and look at the next image, and as we look at the next image, we notice something about it. We are getting to the point where the original image, which we may not have found attractive, is disappearing, and what we are now beginning to see emerge from this completely conceptual collage-making process of repeating this pattern of reducing things and arranging them on the page, we see that what is emerging on the page is this beautiful object that begins to look more and more like a fern. This is the famous *Barnsley's fern*, and if you just looked at that image, this final image, you would find—look at the image; if somebody gave you that image, you would say, "Oh, I'm looking at a fern," but you can see, because we have seen the intermediate steps, that in fact, it didn't come from a fern at all, that that beautiful fern is actually a figment of the imaginations of our minds, and it's created by a very simple process that is just iterated.

In fact then, it looks so natural to us, this final picture of a fern looks so natural to us that it does make us wonder whether nature herself, in creating the natural world, uses processes that are similar to the processes that we used here, namely, iterating a very simple process, and just doing it again, and again, and again.

Let's return to those beautiful, natural images that we enjoyed at the very beginning of this lecture. Look at these beautiful things. Look at this moonscape, these beautiful mountain scenes and snow. Aren't these beautiful images? Now, I have to confess that nature was never involved in the creation of these images. These images were created by the minds of a mathematical process, of which the Barnsley's fern construction was an example. Creating things by an iterative process allows us to capture objects that look so natural that it makes us believe that nature created them that way.

In fact, you can take any picture you want, any photograph or any natural scene, and you can create a fractal approximation to that scene, somewhat like how Barnsley's fern was created. It is sometimes used even in data transmission. Because, for example, if you wanted to transmit the image of that fern, all you would need to do is to write down the equations of where to put those four copied images in the collage process. Just write them out in longhand, send them in an envelope, or type them in a computer email, and you could send just those few sentences that you see here on the screen. Then, the person could take them and create the entire image of the Barnsley fern. This is, therefore, a thought-provoking concept.

One way in which these fractals that we talked about were created was this method of the collage iterative scheme, but there's another method of creating fractals that involves randomness. I'm going to show you how to create a specific picture by a random process. This is kind of an interesting concept, of how you can create something where you know what the final answer is going to be, and yet, the creation of it is completely random. Let's go ahead and see if we can do this.

What we're going to do is revisit the Sierpinski triangle. Now, you remember the Sierpinski triangle. This was this triangle where, if you had the top part, the left part, and the right part, each of those parts were identical to the whole triangle. That was the Sierpinski triangle, but even if you don't remember it, we're going to reconstruct it here, now. We'll start with just three points that make the vertices of a triangle. There you see them, one, two, three, and we will label them 1, 2, 3.

Then, we will take a die, or something that can give us a random number, and we will roll the die. Except let's suppose that it only has the numbers one, two, and three on it. You roll the die, and whatever number you come up with, you move halfway toward that numbered vertex.

Let's just start somewhere. Let's start at the vertex number one, roll the die, and then whatever comes up, we will go halfway toward that number. So let's suppose the first thing we roll is the number two. Then, we go halfway toward that vertex, right along the edge, and there we are, right at the halfway point of the edge between one and two.

Now, we roll the die again. We could come up with one, two, or three, but let's suppose we come up with two again. Then, we go halfway toward the number two again. You see? All I'm doing is moving points, and as I move, and make a dot on the paper. I started with a dot at point one, and now I have a dot at the midway point of that edge, and I have a dot that's actually at the point of three-quarters of the way toward two, and now I roll the die again. Suppose I come up with a three. I move halfway toward the vertex number three, and I put a dot. Now, I roll the die again, and suppose in this case, I came up with the number one. Then, I move halfway toward the vertex number one. Each time I'm doing this, I put a dot down, and what I'm going to do is just do this process many times.

Now, we have only done it perhaps five times here—here, you can see just five times—but I'm going to run a computer program that will show you what happens as we iterate this process hundreds of times. Here, you can see on your screen this process being created. As you see, it's kind of a fuzzy haze, but the fuzziness gets more and more crisp. You can see that as it gets more and more crisp, it begins to look more and more like the Sierpinski triangle. In fact, if you ran this program forever, you would find that you would get something that was indistinguishable from the Sierpinski triangle. Technically speaking, what you get is that every point on the Sierpinski triangle is approached arbitrarily closely by the points that you create in this random method.

Well, the first thing it is sort of interesting to see is that the Sierpinski triangle is emerging from just this random collection of motions of a dot. But let's try to understand why it is that, in fact, we are required to get this particular image by this random process. Let's see if we can understand this.

The first point to understand is that if you start at the point on the bottom left-hand corner, that is a point in the Sierpinski triangle. Because remember that the Sierpinski triangle started with a triangle, and all of the edges of the original triangle are in the final Sierpinski triangle that we saw in the last lecture.

That point is there, then. When we go halfway toward the point number two, suppose we took the entire Sierpinski triangle, every point in the whole Sierpinski triangle, and we just moved all the points halfway toward the apex of the triangle. Well, remember, the top third of the triangle, the one that goes to the halfway line of the triangle, is identical to the whole Sierpinski triangle, but smaller. If we move up just exactly toward that half, we will see that what we're doing is just reconstructing the big Sierpinski triangle, only at half scale, by multiplying by half and moving toward two.

One thing, at least, that we can be confident of, then, is that when we take that first step of moving toward number two, at least we're still on the Sierpinski triangle. Now, when we shook another two, we took the triangle, that's the top half, and we made it into the little triangle that was just a quarter, the top quarter triangle, the one that is at the very apex. That is three-quarters of the way up. It is a quarter in height, compared to the original triangle, so that the entire Sierpinski triangle, now, after two iterations, would just fill out that top triangle. In particular, the thing I want you to focus on now is that at least the point that we're moving is another point in the Sierpinski triangle.

Now, we will say it just once or twice more and then get the idea: That when we shook a three—the whole Sierpinski triangle, that if you took any point in the Sierpinski triangle and you went halfway toward three, you're going to get another point in the Sierpinski triangle. So that in particular, wherever we are—in this case, we were on the edge, between one and two, three-quarters of the way toward two, we went halfway toward number three, and so, we're at another point in the Sierpinski triangle. Then we circled one; well, the whole Sierpinski triangle goes toward one, so we're still on the Sierpinski triangle, and so on.

So far, then, what we can conclude is that when we roll these dice, and we keep moving the points, at least we're staying in the Sierpinski triangle. For example, we never land in one of the open spaces. Right? We never land inside that big triangle in the middle, because that's not halfway toward the point one, two, or three, starting at any point of the Sierpinski triangle. It is also in none of the other gaps. We never land in a gap.

That's the first thing. If you were literally going to use the pointillism method in drawing the Sierpinski triangle, that is, picking points in drawing the triangle, what you would be to do is make sure that your pen landed in every single place where there was a point on the Sierpinski triangle. Right? That's the way you would have to draw it. You would have to draw it point by point by point by point by point. Well, so far we've only got five points. It doesn't look very much like we're getting all the points in the Sierpinski triangle, so that what I need to convince you of is that in this random process of going halfway, halfway, halfway, we will get as close as we want to get to any given point in the Sierpinski triangle. Let me explain what I mean by that.

First of all, is it possible that after we throw the die infinitely many times, would it be possible that we never got into this little corner way down by number three, the one that's a quarter as long as the bottom side? That piece of the Sierpinski triangle? In fact, it is one-ninth of the Sierpinski triangle, that bottom corner. Is it possible that we would never get there?

Well, I claim that we would always arrive there sometime, and the reason is this. If you roll dice infinitely many times, one thing that is going to happen for sure is that eventually, you are going to roll two threes. What happens when you roll two threes? No matter where you were when you bounced around, bounced around—remember that in ten million different iterations, you are somewhere in the Sierpinski triangle. What happens when you roll two threes? Well, wherever you are in whole triangle, whenever you roll one three, you are down in this bottom right-hand third of the Sierpinski triangle. If the next one is another three, then you are definitely in that bottom ninth of the Sierpinski triangle. Whenever you throw two threes in a row, you are definitely down in that little area of the triangle.

Well, let's take a smaller part of the Sierpinski triangle, for example, this little 27th of the triangle. Let's see that it's possible, or that it's necessary, through probability, that when we roll the dice infinitely often, we're going to get at least that close to inside that little part of the Sierpinski triangle, because every time, in order to get to that point, it's a point that you would get to if you were anywhere in the triangle, and you rolled the following sequence of numbers. If you first rolled a two, and then a three, and then a one, wherever you were in the whole Sierpinski triangle, when you rolled a two, you got up into this corner. When you rolled a three, then it became this ninth of the triangle. You have to be in there after rolling first a two, and then a three. Then, when you roll a one, you have to be in this particular 27th. You see? No matter how small a piece of the Sierpinski triangle I choose, there's some sequence of numbers that will force me to get there no matter where I started, but if we're rolling the dice infinitely often, any particular finite sequence of numbers is sure to come up, in fact, infinitely often. Thus, we're putting a dot close to in every single point of the Sierpinski triangle.

That is an interesting concept, I think, the idea that you can create this Sierpinski triangle from a random game. This is called the Chaos Game. By the way, it turns out that it is not even necessary to start with a point in the Sierpinski triangle. If you just start anywhere the world, and you start going halfway toward these numbers, the set of points that is approached infinitely will be the points on the Sierpinski triangle. That's a thought question for you, even though the points you actually draw may not be on the Sierpinski triangle.

This, then, is a way of combining a definite result with a random process, and what I'd like to do for the remainder of this lecture is to talk about an idea of how we can create ideas. The purpose of this whole collection of lectures really has two purposes. One is to show some neat math ideas, but the second purpose of the course is to say: Can we learn methods of thinking that can be applicable in our daily lives, or whatever is that we do, completely outside of mathematics?

I think that this next topic illustrates a strategy for developing ideas that is potent and applicable everywhere. Here's the idea. If we find a situation that is beyond our understanding, a good way to try to approach that situation and to develop concepts that can make us understand it better, is to look at some simple things that we understand well, see a pattern in those simple things, really understand the simple things very deeply, and then use that pattern and apply it to a more complicated situation.

Let me illustrate it here in the following setting. When we have an object like the Sierpinski triangle, that has this sort of infinite complexity to it, literally infinite complexity, with this self-similarity property, one thing that we can look at is that we can say, "Well, it doesn't seem to have the concept of dimension that we are used to." In other words, we really don't want to call it one-dimensional, because isn't just like a line, and really don't want to call it two-dimensional, because it really isn't filled in. Somehow, then, it's got a kind of fuzzy sense to it, and one thing that we might want to do is to take that fuzzy sense and build a concept that would capture that fuzziness, and we could call it something related to "dimension," see? It's something that would capture how thick it is, but it's got to be a slightly different concept, because this realm of infinite complexity is not an object that we are familiar with.

Let's go ahead and try to use this strategy for building a new concept, then. We're going to define a new concept of dimension that would apply usefully to this new idea. The way we do it is to think about familiar objects. Let's start here with a very familiar object.

Here we go. We have a square. Now, the reason we start with something like a square is because a square is something we are familiar enough with to know what its dimension is. Its dimension is two. The reason its dimension is two is because that's what we have been brought up to believe it is. That is, so to speak, the familiar, the norm, of our experience, which tells us that this is dimension two. This is the foundation, the familiar upon which we're going to build the unfamiliar, and we notice something about this square. If we wanted to construct a square that's twice as big, that is, twice as big lengthwise and widthwise, we would need four copies of a square this size to do it. Likewise, if we wanted to create a square that was three times as big, that is, we put it in a copy machine, and entered "300 percent," we would need nine copies of this size of square to make it. One, two, three, four, five, six, seven, eight, nine.

Let's make another observation. So far, all we are doing is looking at familiar objects. We're going to do something with that, and generalize about that information so that we can apply it to the Sierpinski triangle, which is a weird world. Let's look at a cube. Suppose we have a cube here. Here's a cube, and suppose we want to make a cube that's twice as big—twice as long, twice as wide, twice as tall—how many cubes would we need? Well, you need $2 \times 2 \times 2$. You need eight cubes.

Likewise—this is a familiar object, a Rubik's cube—if you have a unit-sized cube that was just one corner, how many little cube switching need to make a 3-by-3 cube? Well, $3 \times 3 \times 3$, 27. Let's see if we can see the pattern.

The pattern is that if we have an object, in this case a square, and we seek to make a square that is two times as big—and we know what we mean by that; we mean two times in all directions—we need four copies. We also notice that—this is an obvious thing—that it is 2^2 equals 4, that the number of copies we need is equal to the number of times bigger that we want it, raised to dimension power. It's two-dimensional, so 2^2 is equal to 4. Let's do it 3-by-3, so that you see the pattern. Okay?

If you want to know how many copies of the unit-sized square to make a square that's three times as big, you take 3, the number of times bigger you want it, raise it to the dimension power, 2, and you get, in this case, nine, $3^2=9$, the number of squares needed to make one that big.

Let's see if this pattern holds for the cubes. Okay? If you take a cube, and you say, "Ah, how many cubes do I need to make something that's twice as big?" We understand what we mean by "twice as big." Well, what you do? You take "twice," the number of times bigger that you want it, and you raise it to the dimension power, 3; 2^3 is 8. This cube correctly predicts this. We will verify it once more. If you have a unit cube, you want to find how many unit cubes are needed to make a unit cube that is three times as big, so three times as big, raise it to the dimension power, 3; 3^3 is 27, the number of cubes needed to make a cube three times.

Okay, we have seen a pattern. Let's first verify the pattern in a simpler case. Suppose you take a unit line segment. How many segments do you need to make a segment that is twice as long? Well, if it's twice as long, you have 2, that's the number of times longer you want it, raised to the dimension power, 1, gives you the length that you need for this line segment that is twice as long. Okay? Or three times as long; 3^1 is 3. It's confirming the idea that the number of times bigger you want it, raised to the dimension power, gives you the number of copies needed to make that bigger object. See? Putting it in English like this then tells us the clue. Let's then use this pattern and apply it to the Sierpinski triangle, and just see what we get.

Take a Sierpinski triangle. There, on the screen, you see a Sierpinski triangle. Now, a Sierpinski triangle has the property that it has parts of it that are the same as the big thing, and so we can count and see how many copies of the Sierpinski triangle are needed to make a Sierpinski triangle that's twice as big. How many copies would we need? Well, look at them. There's the one on the lower left, the one on the lower right, and the one on the top middle. If we want to, then, make a Sierpinski triangle that is twice as big, we need three copies.

Let's then follow the pattern that we discovered in a square and the cube, and see what we get then. It says that if you want to make something twice as big, you raise it to the dimension power to get the number of copies you need. In this case, we have the little equation 2^d, with d as the dimension—we don't know what we're going to call the dimension, but we are defining it. d is that number which, when you take 2 and raise it to the d power, you get 3, $2^d=3$. You see? We are defining a new concept by following a pattern that we saw from familiar objects.

Now we have an interesting thing, though, because we have this equation: 2 to the dimension of the Sierpinski triangle, what we want to call the dimension of the Sierpinski triangle. We are in control; we are defining it, so that 2, the number of times bigger, to the dimension of the Sierpinski triangle, is equal to 3, $2^d=3$. What would that dimension be? Well, we actually just take out a calculator and plug in numbers to see what number d do you have to put to get 2^d, to get 3? Well, it's not the number 1, because 2^1 is 2. It's not the number 2, because 2^2 is 4. 2 is too big, 1 is too small. You know, it's sort of like the three bears, "too hot," "too cold." It's got to be somewhere in between, and if you actually compute it, you see that d is equal to 1.58496, and so on. It is a fractional number. It is exactly equal to, in fact, the log of 3 divided by the log of 2, but we don't need to know that. All we need to know is that the number is between the numbers 1 and 2.

We've therefore developed a concept that gave us an interesting answer. It gave us a measure of a feature, of an object in the world, in this case, a conceptual object, but we got to that feature by looking at the familiar, and then applying the concepts that we learned, the pattern that we saw in familiar objects, and applying it in this unfamiliar realm. This, I think, is a good illustration of a strategy of thinking. As a homework exercise, you could compute the fractal dimension of the Menger sponge. Remember, the Menger sponge has 20 copies that make a sponge that is three times as big as one copy.

I hope that you have enjoyed this excursion both into fractals and into the concept how to develop ideas.

Lecture Nineteen
Chance Surprises—Measuring Uncertainty
Michael Starbird, Ph.D.

Scope:

Many, if not most, significant events in our lives arise from coincidence, randomness, and uncertainty. We stumble upon people who will eventually become important in our lives; we accidentally find intriguing opportunities; and we fall into a profession or lifestyle. Nothing is more fundamental than chance. The uncertain and the unknown are not forbidding territories into which we dare not tread. They, too, can be organized and understood. We can construct a means to measure the possibilities for an undetermined future. Quantifying the likelihoods of various uncertain possibilities is an impressively grand idea. How can we sensibly measure what we admit we do not know?

We begin with some scenarios in which chance dictates unexpected outcomes. When a circumstance appears surprising, that is a signal that we must re-hone our intuition. We proceed to develop conceptual tools that help us accurately measure the possibilities of an uncertain future.

Outline

I. A major intellectual accomplishment of human culture is that we can, in a sense, measure chance. We can assign a value to the likelihood of uncertain events.

II. To understand chance, we'll use a method that we have used earlier, that is, to examine simple cases and apply our findings to more complicated situations.

    A. We start with a display of pennies balanced on their edges. When the table is hit, most people would probably guess that about half the pennies will fall heads up and about half will fall tails up.

        1. The results of this experiment, however, may cause us to reevaluate our intuition about this situation. In fact, more pennies fall heads up than tails up.

        2. We can perform a similar experiment with spinning pennies. Surprisingly, spinning pennies will land tails up about 70 percent of the time.

B. These counterintuitive results prompt us to define what we mean by the idea of measuring chance.
  1. The basic measure of likelihood is calculated by dividing the number of times a particular outcome occurs by the total number of attempts to reach that outcome.
  2. The resulting fraction is called the *relative frequency*.
  3. The relative frequency approaches the probability with increasing accuracy as the number of trials increases. This concept is called the *law of large numbers*.

C. Let's look at an even simpler example of probability than the pennies, namely, rolling a die.
  1. Without even rolling the die, we know that the probability of rolling a 3, for example, is $\frac{1}{6}$, because the number of equally likely outcomes of rolling a die is six, and 3 is one of those outcomes.
  2. A slightly more complicated situation is rolling two dice. Are the probabilities equally likely of rolling, for example, a 7 and an 11?
  3. We could roll a 7 in six different ways; thus, the probability of rolling a 7 is $\frac{6}{36}$, or $\frac{1}{6}$. The probability of rolling an 11 is $\frac{2}{36}$, or $\frac{1}{18}$.

D. Another example comes from a college reunion.
  1. One classmate says to another that she has two children, the older of which is a boy. Before she can identify the gender of the other child, she chokes on an hors d'oeuvre and falls to the floor. What is the probability that she has two boys?
  2. Another classmate says that she has two children, one of which is a boy. Before she can identify the gender of the other child, she, too, chokes on an hors d'oeuvre and falls to the floor. What is the probability that she has two boys?
  3. In the first case, when the woman says that she has two children, the number of equally likely possibilities for the combination of gender of her children is four. When she says that the older child is a boy, two of these

possibilities are eliminated. The probability that she has two boys is $\frac{1}{2}$.

4. In the second case, only one of the equally likely possibilities is eliminated when the woman says that she has a boy. The probability that she has two boys, then, is $\frac{1}{3}$.
5. These examples illustrate the idea that in determining probability, we are trying to reason out a measure of the unknown.

III. If we have two dice, what is the probability of rolling two 6s?
   A. Looking at the chart of equally likely outcomes, we see that the probability is $\frac{1}{36}$.
   B. We can also multiply the probability of rolling a 6 on each die to arrive at the probability that we will roll two 6s: $\frac{1}{6} \times \frac{1}{6} = \frac{1}{36}$.
   C. The probability that two independent events will occur is equal to the probability that the first will occur multiplied by the probability that the second will occur.
   D. For example, in flipping a coin and rolling a die, the probability that the coin will land on heads and the die will land on 6 is $\frac{1}{2} \times \frac{1}{6} = \frac{1}{12}$.
   E. Notice that the distinction between *and* and *or* in these questions is important.
   F. Complex DNA fingerprinting can be interpreted with these same principles.
      1. The probability of having a long string of gene pairs match exactly for two different people is the probability that the first pair matches multiplied by the probability that the second pair matches and so on.
      2. We can understand this complicated scientific process by applying our knowledge of simpler situations of probability.

Suggested Reading:

Burger, Edward B. and Michael Starbird, *The Heart of Mathematics: An invitation to effective thinking*, Key College Publishing, Sections 7.1, "Chance Surprises: Some scenarios involving chance that confound our intuition"; 7.2, "Predicting the Future in an Uncertain World: How to measure uncertainty through the idea of probability"; and 7.4, "Down for the Count: Systematically counting all possible outcomes."

Questions to Consider:
1. Balance pennies on their edges on a table. Then slam your hand on the table, causing the pennies to topple over. Perform this experiment until 100 pennies have fallen. Record how many pennies landed heads up and tails up. What percentage of the pennies landed heads up? Next, spin pennies on their edges 100 times and see how they land. Record your results. What percentage of the pennies landed heads up? Finally, flip a penny 100 times and record how many pennies landed heads up and tails up. What percentage of the pennies landed heads up?
2. Take a shuffled deck of cards and remove cards from the top of the deck in pairs. For each pair where at least one of the cards is black, record whether they are both black or whether one is black and one is red. After you have done five pairs, reshuffle the deck and repeat until you have recorded 100 cases where at least one card of the pair was black. What percentage of those pairs had both cards black?

Lecture Nineteen—Transcript
Chance Surprises—Measuring Uncertainty

Welcome to Lecture 19: Chance Surprises—Measuring Uncertainty. Before we embark on this new topic, let's take a minute to see where we are in the whole course. The whole course grouped mathematical topics into issues concerning number, geometry, and uncertainty. With the previous lecture on fractals, we ended our section of geometrical themes. So we now embark on our last set of five lectures, and look at how we measure uncertainty. In these lectures we will confront the formidable challenge of predicting the future. The challenge is to measure the uncertain and the unknown in some meaningful way. I'll begin with two lectures that introduce the basics of probability through some surprising examples. Ed will follow with three lectures involving different aspects of chance: expectations, randomness, and coincidence.

We start this lecture with the basic question: How can we measure chance? That question is certainly worth a great deal of thought, because many of the most important parts of life involve chance—the people whom we meet and who become our friends and spouses, the teachers who influence us, the ideas we find attractive at susceptible moments, the particular jobs we fall into, our co-workers and neighbors, and the societal challenges that happen to confront us at our particular moment in history. All these crucial aspects of life involve randomness and chance to a very large degree. If we look at our real life history, we might well say that essentially all the most important forks in our life's path occurred owing to chance, randomness, and coincidence.

So a truly great intellectual accomplishment is the idea that we can consider an uncertain event, one where we know we don't know whether or not it will happen, and somehow meaningfully measure the likelihood of its happening. We will see how to put a number between 0 (it definitely will not happen) and 1 (it definitely will happen) on future events that may or may not occur. To me, the whole idea of associating specific measurement, that is actual numbers, to future events that are not determined, is amazing. In other words, we know for sure we don't know what's going to happen, and yet somehow we can measure the likelihood in a very practical way that makes a real difference in our lives. When we hear there is a 90 percent chance of rain, we take an umbrella.

What does it mean to put a specific number to the likelihood that something will happen, when what we really know is that we don't know whether it will happen or not? The strategy by which we will take this fundamental conundrum of life, namely chance and uncertainty, and make sense of it and put some structure to it will be a strategy that we have used several times before. That is, we'll use the method of looking at simple cases and understanding them so clearly that we can apply our insights to more complicated situations. This strategy of focusing on simple cases is a doable activity that allows us to take real action in the domain of the familiar to pull ourselves up by our bootstraps to be able to deal with far more complex situations. That strategy is a great way to understand a complex and difficult issue. And what could be more complex and difficult than what we are about to attempt, namely, trying to measure the uncertain and the unknown, or put more provocatively, in some practical sense to predict the future. What could be more important than finding organization even in chance?

As Boethius said it, "Chance, too, which seems to rush along with slack reins, is bridled and governed by law."

What we're going to do is start with this interesting experiment. I hope that you see on this table, this amazing situation, that we have a whole bunch of pennies here, and these pennies are all standing on their edges, and believe me, this is not the easiest thing to get a bunch of pennies to stand on their edges, but you can try it at home. It's hard, though. Here we have all these pennies standing on edge, and they are all delicately balanced. Now, what I am going to do in a minute is to hit the table, and they will all fall over. Some will land with their heads facing up, and some will land with tails facing up. These are all pennies, so some of them, heads up, some of them, tails up.

Now, you might have a guess about what's going to happen here, and some of you may want to bet in the privacy of your own homes; if there's someone there, you might want to wager. Are there going to be more heads? Are there going be more tails?

This is a question about probability. Probability, you see, is the way of measuring the uncertain and the unknown. So what we're going to do is first do this experiment, and then we're going to talk about what the experiment means. Let's go ahead, then, and try. I am going to hit the table, and is everybody ready? Do you have your guess for heads or tails? Here we go.

Okay, well, at least they all fell down. Now, let's see. I'm going to gather all the tails on this side, and all of the heads on this side. Now, let's see. I'm looking here. I see quite a few heads. Let's see, there's a tail, there's a tail, there's a tail, there's a tail, here are some tails, here are heads, and I want to be pretty honest about this, so that I don't cheat. I could probably cheat, and you wouldn't notice, but here's a tail. Let's see, these are tails, this is a tail, and this is a tail, and all the rest of these are heads.

Now, if we look at these—let's just look at them, and you might notice that just by looking at the size, that there are quite a few more heads than there were tails. Now, this is an interesting fact. First of all, if you've made a wager, now is the time to collect on your bet if you bet that heads were the thing to have. This is an interesting thing. We had an original guess about what would have happened. I would think most people would say that the chances were about 50-50 that the coins would land heads or tails. Here we have some evidence, though, that tends to indicate that maybe that's not the case. In fact, there are really quite a few more heads than there are tails, maybe even almost twice as many heads as tails. This makes us reevaluate, then, our understanding of this particular sort of trivial kind of situation, and yet one that will illustrate a concept of measuring probability.

I'm going to do one more thing before we talk about it in some more detail, and that is that I am going to spin some pennies. Here's what I'm going to do. I'm going to spin a penny, and then, it will just land, so it spins for a while. First of all, I hope you're impressed that it lands, and then, it lands. That landed a "head." Then, here we go. We'll spin it again, and we'll see whether it lands heads or tails. That time, it landed a tail.

Now, I'm not going to take the time to spin it a lot of times. But I've done this already in the past, and when I did it, I spun 100 pennies; and when I spun 100 pennies, 70 times they landed tail up. Seventy times they landed tail up. So this is sort of an interesting fact about pennies. I'm going to propose that you can, in fact, never buy a dinner again, you can get your companion, whomever you go to dinner with, to pay for dinner, and this is the way you do it.

You see, what you'll do is that you go to dinner, and you're trying to decide who's going to pay for the check, and so you say, "Well, okay, why don't we just do just a random process with a penny, and then if it comes up heads, you win, and if it comes up tails, I win, but you go ahead and choose."

Now, if the person chooses, say, heads, then you say, "Well, instead of just flipping a coin, which is so mundane and boring, why don't we just spin them? We will spin a lot of them, so it is more random." You spin a bunch of pennies, and you'll find that almost always, you will win, because about 70 percent of the time, it comes up tails. If they bet tails, then you say, "Oh, let's not just do this coin flipping thing. Let's just balance the coins on edge." And then you hit the table, and you'll find that almost always, the head side comes up more frequently.

Now that you have learned how never to pay for dinner again, let's go back to the question of what we are going to mean by the concept of measuring an uncertain situation. Well, we have a concept of that measure. The measure of something uncertain is that, in the cases of pennies, when we had pennies up on edge, if we had a lot of pennies—as we did here, we had a lot of pennies—if we do the experiment we're talking about, in this case, balancing them on edge and hitting the table, a certain fraction of pennies will land heads up, and a certain fraction will land tails up. That fraction is really what we mean by the likelihood, the chance of a penny landing heads up.

You see, if you just take one penny, and you balance one penny, and you hit the table, and you say, "What's the chance that it will come up heads?" Well, either it's going to come up heads, or it is going to come up tails, if you just do one experiment.

What does it mean to measure the likelihood? Well, one concept of this is that you imagine doing the experiment many, many times—like thousands of times—and suppose you do this experiment thousands of times, and you discover, doing it thousands of times, that maybe 65 percent of the time, heads comes up, rather than tails. Then you can say that the probability of a coin coming up heads if you do one experiment is 65 percent for heads. That's the probability that it will come up heads.

The measure of an uncertain event means that you're trying to measure how often the outcome that you're talking about—landing heads if it's balanced—comes up, versus how many times you tried the experiment. The number of a certain outcome, then, divided by the total number of trials is a measure of probability. In fact, the number that you get, the probability that you get, is, if you do a lot of experiments, the actual measurement of the fraction will tend toward that number that you call the probability, and that is called the *relative frequency* of coming up heads versus tails, and the number of heads, over all outcomes, as the relative frequency of those. It has sort of a fancy name to say that if you do an experiment a lot of times, then the fraction of times that it comes out with a certain outcome is equal to the probability. That is called the *law of large numbers*.

We have seen an example of pennies. The thing about this pennies experiment is that it was a little bit surprising. Right? Because we guessed 50-50, and it came out to be different from that. Let's look at a situation that is simpler, so that we can actually develop the concept of probability more clearly. A situation that is simpler is a case where you take an object like a die—dice, by the way, are great sources for understanding probability, because a die is a completely symmetrical object. It has six faces, as you know, and if we number the faces "1, 2, 3, 4, 5, 6," then, if the die is completely evenly balanced, and we throw the die randomly, the chance of it landing one side up is exactly the same as the chance of landing any other side up.

This really gives us the concept of probability in the abstract. Instead of actually rolling the die, before we even roll the die, we can see what we would mean by saying the probability of rolling a die and coming up a 3, we would know what that means. It means that you've got six equally likely things that could happen, and one of them is a 3, the probability of rolling a 3 is one out of six possible things that could happen. You see? There are six equally likely ones, and rolling, say, a 4, is one out of six. You say that the probability of rolling a 6 is one out of six, or rolling a 4 is one out of six.

One of the most fundamental concepts of measuring uncertainty of future situations is the following: If you can say, "I'm going to perform an experiment where there are equally likely things that might happen," then, you would say, "If there are six equally likely things that might happen, then the probability of one of them happening is one out of six." Very straightforward. We have to be cautious, and that's why I did the penny experiment. You might guess that there was a 50-50 chance that when you balance the pennies on edge, that there's a 50 percent chance of whether it lands heads, and a 50 percent chance of whether it land tails, but that's not the case, as we saw. You therefore have to be cautious, but in the abstract, a die that's evenly balanced we'll say that it has an equally likely chance to land on any of the six sides.

Now, let's move on and talk about a slightly more complicated situation, and that is where we have two dice. If you have two dice, and you do an experiment, this is a place where you can easily go wrong in your analysis. When you roll two dice, whatever you roll, one die is going to be a number between 1 and 6, and the other one is going to be a number between 1 and 6, so that the total will be some number between 2 and 12. Right? Two and 12, those are the possible outcomes when you roll two dice.

You might say, well, okay, it could come out a 7, right? It could also come out an 11, or it could come out a 3, or it could come out a 10. You might say, then, well, okay, since it could come out a 7, or it could come out an 11, you might say, the probability of it coming out a 7 is the same as coming out an 11, and is equal to, like the probability of coming out a 7, since there are 11 numbers that it could come out—2, 3, 4, 5, 6, 7, 8, 9, 10, 11, or 12—there are 11 possible outcomes, you might say then the probability of getting a 7 is one possible outcome out of the 11.

There's a fallacy to that reasoning, though, and the fallacy is something that you intuitively know, namely, that it is a lot more probable to get a 7 than it is to get a 2, or an 11. Why is that? Let's see if we can understand what it means to measure the probability of getting a 7.

What we need to do is figure out what the equally likely outcomes that could occur are when we do the experiment of rolling two dice. What we do is that we have a chart, as you see, that illustrates the 36 equally likely possible things that could happen. It is important to view these two dice differently, so that is why we have differently colored dice. This is a green one; this is a yellow die. These two dice, being differently colored—we see that this die, when you roll both of them, the green die could come up any number from 1 to 6. Likewise, the yellow die could come up any number from 1 to 6. The probability of any particular outcome, then, such as green-3 or yellow-4, is one out of the 36 equally likely possible outcomes.

Now, let's talk about rolling a 7. What is the probability of rolling a 7? Well, now we have to think it through. Rolling a 7 can be done in one of many different ways. It could be done by rolling a 6 on the green die, and a 1 on the yellow one; or a 5 on the green and a 2 on the yellow; 4 on the green, 3—you see all the different possibilities. In fact, the green die could have any of the numbers between 1 and 6, and then the yellow die would have its number determined, and it would be some number like a 6 and a 1; or a 5 and a 2; or a 4 and a 3, and so on. There are six different outcomes of the two dice, how they land, that all add up to 7. If you wanted to talk about the probability, then, of getting a 7, what you need to actually measure is that there are six different outcomes of the 36 possible equally likely outcomes that result in numbers that add up to 7. The probability of rolling a 7 is those six equally likely possible things that could happen that give a 7, out of the 36 actual equally likely possibilities that can happen when you roll two dice. The probability, then, of throwing a 7 is exactly 1 in 6. You should think of it, though, as 6 out of 36.

Now, let's think about rolling an 11. Rolling an 11 is a different kettle of fish, because to roll an 11, you don't have 6 possibilities. The first die has to be either a 5 or a 6. So the green die could be a 6 and the yellow one could be a 5—that's one possibility—or the green one could be a 5 and the yellow one could be a 6. There are no other possible combinations of those two dice though that could give you an 11, so there are only 2 chances out of 36 of throwing an 11.

That's where we get the concept of probability for equally likely events. Let me give you an example of a scenario that happened to come up when I was at a reunion of my college class. I was attending a reunion. This is a true story. I was attending a reunion, and it was the 30th reunion of my college class. I was visiting these people, and we were having a little cocktail party. Hors d'oeuvres were being served, and this woman came up to me, and of course, she was very old. I don't know who she was, but I certainly didn't know her. She was much older than I.

Anyway, she was in my class, so she came up, and she said, "I have two children." I said, "Oh, isn't that interesting? Is your older child a boy?" She said, "Yes," and then right then, she choked on an hors d'oeuvres, and just fell down on the ground, choking, unable to speak further. She was there, turning blue. It was really quite a scene. And so, I was sitting there, and I was thinking to myself, "What is the probability that she has two boys?"

You see? Because I knew the older one was a boy, but I was thinking, "What's the probability that she has two boys?" She was sitting there choking, and I went over to the next person at the reunion, and there was another old woman, and she came up to me, and she said, "I have two children," and I said, "Oh, isn't that interesting?" I said, "Do you have a boy?" She said, "Yes, I have a boy," and then she choked. She choked, and fell on the ground, and started to turn blue. Now, they're both on the ground turning blue, and I'm thinking to myself, "What is the probability that she has two boys?"

Now, of course, just listening to this, it sounded like the same conversation, didn't it, in these two cases? Of course, I can also tell that you are appreciating my compassion for the suffering of my fellow students that were choking. Anyway, I'm asking myself these probability questions: What are the chances that there are two boys?

Well, let's relive this. There were two different scenarios, and they were slightly different, so I have to be a little careful, very careful of what I said. In the first one, I asked, "Is your older child a boy?" She said, "Yes," and then choked.

In the second one, I said, "Is one of your children a boy? Do you have a boy?" She said, "Yes," and then choked.

Let's see if we can understand exactly what happened in these cases. In the first case, when she said, "I have two children," what came to my mind? As soon she said, "I have two children," I knew that one of the following four equally likely events had occurred. Either she had an older boy and a younger boy—see? Older boy, younger boy, so this is older, and younger; or she has an older boy—we will put the older ones on this side, and the younger ones on this side—older boy, younger boy; older boy, younger girl. That could happen with two children. Older girl, younger boy. That could happen. See, now there are three possibilities. Or, older girl, younger girl.

One of these four possible situations could occur, and these are equally likely to happen. You see? Well, now, that was when she had just said, "I have two children." As soon she said, "I have two children," in my mind, I'm thinking, "Here are four equally likely scenarios that might happen." Now, I ask the question, "Is your older child a boy?" Well, she said, "Yes." Well, when she said, "Yes," then I said, "Well, either she has this situation or this situation, but she doesn't have two girls," so we throw those away. And she doesn't have an older girl and a younger boy, so I put that away. The effect was that after she said, "My older child is a boy," and she affirmed that was true, then the two equally likely possibilities were that she had two boys, or she had an older boy and younger girl. Those were the only two things left, and so the probability of having two boys was one-half.

Now, let's think about the other scenario. The other scenario was that she came up to me, and she said, "I have two children," so here they are, she had two children. Once again, I was thinking to myself, as soon she said, "I have two children," they were either boy-boy, boy-girl, girl-boy, or girl-girl. Okay? Those were equally likely things to happen. Then, I said, "Do you have a boy?"—meaning, "Do you have at least one boy?" She said, "Yes, I have at least one boy." So then, I said, "Well, then, I know for sure that it's not this situation with the two girls." I threw those out, and consequently, I have left in my mind as the equally likely things that might have happened in her life either older boy-younger boy, older boy-younger girl, or older girl-younger boy. What is the probability, then, of having two boys? The probability is one out of three. You see, then, it was a different probability in the case where she just said, "I have a boy," compared to saying, "My oldest child is a boy."

Now, to a lot of people, this, by the way, is counterintuitive. The reason it is counterintuitive is because if a person says, "I have a boy," then, you might argue the following. You might say, "Well, look. If I have two children, and I have a boy, then, the other one is either a boy or a girl, so it sounds like it's 50-50." You see, though, this illustration shows that, in fact, the probability, if you say, "I have a boy," meaning "I have at least one boy," is only one out of three.

This was actually in the Marilyn vos Savant's column of the newspaper. This was a raging controversy for several months, when people were talking about the scenario of the boys and girls. It illustrates the concept, though, that we are trying to develop, that in thinking about probability, what we're looking at is that we're trying to reason through a measure of the uncertain and unknown by taking very simple situations like the dice one, where we know what's going on. There's an equally likely chance of any of the six sides coming up, and it developed for us a concept, and the concept was that if we can enumerate the equally likely possible outcomes of an experiment, then, in fact, we can say that the probability of a successful outcome is just counting which of those equally likely ones are the ones we're interested in, divided by the total number of equally likely outcomes.

That is a principle that we can now apply in many different settings. Let's go ahead and see if we can deepen our understanding of this concept. If you have two dice, and you're asking the question, "What is the probability of getting two 6s when I throw two dice?" What is the probability of getting two 6s? Well, first of all, looking at the chart of the 36 equally likely outcomes of the two dice, we see that the chance is 1 out of 36. Let's think about it in a different way, though, in order to understand it better. One way to understand it better is this: The probability of one die coming out a 6 is one-sixth. Right? If you roll a die, what is its probability of landing as a 6? One out of six. Likewise, the probability of the other die rolling a 6 is one of six.

Now, notice that the product of the probabilities, one-sixth times one-sixth is 1 out of 36, $\frac{1}{6} \times \frac{1}{6} = \frac{1}{36}$. If you're doing two independent events, such as rolling two dice, then the probability of both of them doing something is the product of the probability of the first one happening and the probability of the second one happening. Now, we understand that because of the chart. You see, if you wanted to list all of the equally likely occurrences when you have two things going on—in this case, a green die being thrown, and in the other case, a yellow die being thrown—if you wanted to enumerate all of the equally likely outcomes of the two things happening at once, what you would do is make a rectangle. The rectangle would have all of the equally likely events of the one die on the one side, and all the equally likely events of the other one on the other side. Then, every point in the chart is "this outcome, this outcome."

Let's do another example here, before we talk some more about it. Suppose we have two things here: a die and a penny. Okay? We're going to roll the die and get a number, and then we're also going to flip a coin. Now, by the way, let me tell you that flipping a coin really is, as you expect, a 50-50 proposition. The probability of it landing a head is 50 percent, just like it is for tails. So, unlike the balancing of a penny or spinning the penny, now, flipping a coin is really a 50-50 job. Okay? If we now want to enumerate the equally likely outcomes of the event of doing two things, of rolling a die and flipping a coin, there are six possible outcomes for the die, and two possible outcomes for the coin, and therefore, there are 12 possible outcomes for the die and the coin. If we wanted to know: What's the probability that we roll a 6 on the die and heads on the coin?" The probability of doing both those things is just one of the squares of those 12 equally likely squares that could happen.

Now, it's very important to make a distinction between the words "and" and "or." Suppose we ask the question: What is the chance of getting a heads **or** a 6?" Well, if you wanted a heads or a 6, you see, that is a totally different proposition. Because if you look at the chart of these 12 equally likely outcomes, you could get a 6 and a head, or a 6 or a tail. So if you are talking about getting a 6 or a head, as soon as you get heads, that's half of them already. Six of the outcomes of these 12 equally likely ones are heads. "Or" and "and" are therefore words that we have to be careful with, and we will see that a little bit later.

In talking about the probabilities of multiplying, of "and," a good example of that in real life has to do with DNA fingerprinting, like when O.J. Simpson was on trial, and I hope we all watched that very carefully. They did this blood analysis, and they said: Well, if you have a strand of DNA, there are little parts of it, genes that come in different flavors called alleles. If you have 20 different ones of these genes that could come each in two different flavors, the chance of a random person having the first allele the same, and the second the same, and the third the same, is the product of the probability of having that first one the same times the probability of the second one the same times the probability of the third one the same.

If you have a long DNA strand, and each little piece is independently created by luck, and if you then match all of these random, independent events together, you multiply the probabilities that all of them are going to occur. The probability, then, of getting exactly that particular collection of alleles is the probability of the first one times the probability of the second one times the probability of the third one, and so on. That's why you can use DNA fingerprinting to identify people with complete precision.

This was another example of using simple examples to develop larger principles.

Lecture Twenty
Door Number Two or Door Number Three?
Michael Starbird, Ph.D.

Scope:

Starting in the 1960s and for many years thereafter, the game show *Let's Make a Deal*® entertained viewers with Monty Hall urging contestants to pick a door. This game show involves a question of chance that surprises people to this day and has been the source of many heated arguments. Here, we will explore the mathematics that prepares us for future game-show stardom and explains a paradoxical example of probability.

Other examples illustrate the concepts of independence of possible outcomes and the insight that often the best way to measure the probability of an event is, instead, to measure the probability that it will not happen. If two independent things must both happen, the chances become slimmer, and the method to learn that probability is to multiply. We'll see why.

Outline

I. This lecture explores a method of thinking that involves looking at the opposite of a question that we are interested in answering.

   A. We'll use this strategy in connection with probability.

   B. We will also use an approach that involves exaggerating a situation to pinpoint its salient feature.

   C. Finally, we will discover that sometimes, intuition will lead us astray.

II. Remember that the whole concept of probability is to put a number to what we don't know; it is a measure of uncertainty.

   A. In the last lecture, we saw that in rolling a die, the probability of rolling a 2, for example, is $\frac{1}{6}$.

   B. We also saw that if we roll a die and flip a coin, the total number of possible outcomes is 12, found by multiplying the numbers of possible outcomes of each event.

   C. Now let's ask, "What is the chance of rolling the die to get 6 and flipping the coin to land on heads?"

1. If we look at our chart, we count seven possible ways to land on heads *or* roll 6. There is only one way to land on heads *and* roll 6.
2. The distinction between *and* and *or* is important. Figuring out *and* probabilities is often easier than determining *or* probabilities.
3. To find an *and* probability, we multiply the two probabilities of each outcome. For example, the probability of rolling a 6 is $\frac{1}{6}$, which we multiply by the probability of landing on heads, which is $\frac{1}{2}$: $\frac{1}{6} \times \frac{1}{2} = \frac{1}{12}$.
4. The *or* probability can be a bit more difficult to determine.

D. To find the *or* probability, we'll look at the opposite of the question.
1. What is the probability of not rolling a 6? The answer is $\frac{5}{6}$.
2. What is the probability of not landing on heads? The answer is $\frac{1}{2}$.
3. We can easily compute the probability of not rolling a 6 and not landing on heads as $\frac{5}{6} \times \frac{1}{2} = \frac{5}{12}$.
4. We have changed a complicated *or* problem into an easier *and* problem.

III. This principle leads us briefly into the history of probability with a 17th-century French gambler named Antoine Gombauld, the Chevalier de Méré.
A. One of Gombauld's favorite dice games, which he often won, was betting that a 6 would appear at least once in four consecutive rolls of a die.
1. What is the probability of getting a 6 in four rolls?

2. This question is somewhat complicated, but we can answer it by looking at the opposite question: What is the probability of not getting a 6 in four rolls of the die?
3. The chance of not getting a 6 on the first roll is $\frac{5}{6}$, which we would multiply by the chance of not getting a 6 on the second roll, $\frac{5}{6}$, and so on. The overall probability of not getting a 6 in four rolls is $\left(\frac{5}{6}\right)^4$, or .4822.... In other words, the chance is slightly less than half that we will not get a 6 in four rolls of the die.
4. The opposite of that is the percentage of times that we will roll at least one 6, which is slightly better than 50 percent, or .517....

B. Later, Gombauld changed the game to betting that he would get at least one pair of 6s in rolling two dice 24 consecutive times. This game, he lost.
1. He wrote a letter to Blaise Pascal asking why he was losing this game. Pascal, in turn, consulted Pierre de Fermat, and their subsequent investigations formed the basis of probability theory.
2. If we do the calculations, we see that the chance of not getting a pair of 6s in one roll is $\frac{35}{36}$. The chance of not getting a pair of 6s in 24 rolls, then, is $\left(\frac{35}{36}\right)^{24}$, or .5085..., slightly above 50 percent.
3. The opposite of that, what Gombauld was betting on, is .491..., or slightly less than 50 percent.

IV. We now move from the 17th century to the 1960s and the TV show *Let's Make a Deal*®.
   A. The host of the show, Monty Hall, would show a contestant three doors, behind one of which was a wonderful prize, such as a car.

B. After the contestant chose a door, Monty Hall would open a different door, revealing a joke prize that the contestant would not be interested in. Then, Monty would offer the contestant the opportunity to change his or her original selection.

C. It seems that the probability of finding a car behind either of the two remaining doors would be the same, but let's stop and think about this problem.
  1. One way to answer the question of whether or not the contestant should switch doors is to perform experiments to simulate the situation. Another method of thinking about this question is to exaggerate the situation.
  2. Suppose we return to the beginning of the game, but this time, Monty Hall shows the contestant six doors. After the choice is made, Monty reveals unwanted prizes behind four of the doors.
  3. Do we still believe that there is an equal chance that the car is behind either of the two remaining doors, or has our intuition about the situation been altered?
  4. Suppose we exaggerate the situation to 100 doors or even 1 billion, and Monty Hall opens all but two of them after the contestant has made a choice. Should the contestant switch doors or stick?
  5. At this point, the contestant is probably thinking about switching, because the chance that he or she would have chosen the correct door in the first place is 1 in a billion. If the contestant was wrong with the first guess—and chances are that the first guess was wrong—the chance of winning the car by switching guesses is 999,999,999 in a billion.

D. Now let's return to the original situation of three doors and see if we can measure the probability of winning there.
  1. The chance that the contestant was wrong with the first guess was $\frac{2}{3}$. Therefore, if he or she switches doors after one prize is revealed, the contestant has a $\frac{2}{3}$ chance of winning the car.

2. If the contestant sticks with the original guess, he or she has only a $\frac{1}{3}$ chance of winning the car.

V. Let's close with an interesting dice game played with an unusual set of dice.

A. One die has 3s on all six sides. One die has 5s on three sides and 1s on three sides. One die has 6s on two sides and 2s on four sides. Finally, one die has 4s on four sides and two blank sides.

B. Each player picks a die and rolls 20 times. Whichever player rolls a higher number the most times wins.

C. The players can analyze the set of dice to see which choice might better their chances of winning. A player might choose the 3 die over the 6 and 2 die, for example, because the 3 die will win 2 out of 3 times against the 6-2 die.

D. By analyzing all the combinations, the second player to choose a die can always select one that will beat that chosen by the first player $\frac{2}{3}$ of the time.

Suggested Reading:

Edward B. Burger and Michael Starbird, *The Heart of Mathematics: An invitation to effective thinking*, Key College Publishing, Sections 7.1, "Chance Surprises: Some scenarios involving chance that confound our intuition"; 7.2, "Predicting the Future in an Uncertain World: How to measure uncertainty through the idea of probability"; and 7.4, "Down for the Count: Systematically counting all possible outcomes."

Questions to Consider:

1. To win the jackpot of a lottery game, you must correctly pick the six numbers selected from the numbers 1 through 40. What is the probability of winning the lottery?

2. From a deck of regular playing cards, remove three—a king and two 2's. Have a friend act as the dealer. The dealer shuffles these three cards and places them face down on a table, side by side, without looking at them. Once the cards are on the table, the dealer peeks under each card so that the location of the king is

known to the dealer, but not to you. Point to a card. The dealer then turns over one of the other two cards to reveal one of the 2's. Stick with your original guess, turn over that card, and record whether you have chosen the king. Have the dealer scramble the cards again and repeat the exact same scenario—again don't switch—and record the result again. Repeat this experiment 100 times (you can get very quick at it). What percentage of the time did you choose the king? Now repeat the whole process 100 more times, but this time switch your guess after the dealer turns over a 2. What percentage of the time did you choose the king this time?

Lecture Twenty—Transcript
Door Number Two or Door Number Three?

During this collection of lectures, we're really amassing a whole collection of strategies by which we can analyze the world in a more exact and penetrating way. One way we are going to illustrate that, in this lecture, is in the concept of looking at the opposite question that we are actually interested in considering. This is often a useful thing. If we are considering how to be good, maybe it's good to think about how to be bad, and then you know that you can avoid that. Thinking about the opposite then is an important strategy, and we will illustrate it with some probability issues.

The second one that I believe is a very important method of looking at the world is to exaggerate things, to take some situation in the world, and then just exaggerate it like mad, and just in your mind, go through the experience of saying, "What would the world be like if it had this completely exaggerated feature?" Because sometimes using that strategy you'll see that the salient feature is going to be emphasized, and so, exaggeration is a terrific strategy.

Then, finally, we're going to see one thing that's more of a cautionary note, and that is that sometimes our intuition will lead us astray. In probability, this is a wonderful setting in which we can see some things that just don't seem right. Probability will measure them, and see what actually happens.

Now, remember that the whole concept of probability is to put a number to what we don't know. It is a measure of our uncertainty. Probability, then, the way that we have talked about it in the previous lecture, is that we said: If we are doing something—for example, rolling a die, which is a very typical thing that we do for probability. We roll the die. If they are trying to measure the probability of it coming up, say, for instance, a 2, we say that there are six equally likely things that it could come up: 1, 2, 3, 4, 5, or 6. And a 2 is one of them, so the probability is one out of six. We saw that last time.

And last time, we also saw that if we do two things, for example, we roll a die and we flip a coin, then there are going to be 12 possible outcomes—because the coin can come up one of two ways, the die can come up six ways; you multiply the two together; and you could make a little chart that has "Dice: 1 through 6" on the top, "Coins:

heads and tails" down the side. You would have 12 spots in there; those are the 12 equally likely things that could come up.

Now, let's ask a question. What is the chance of getting a heads or getting a 6? First of all, these are boring questions, but the reason we do boring things goes back to one of the previous life lessons. We look at simple situations; we understand the simple ones, and then we can apply them to really interesting things that we actually want to know about. For now, though, let's just ask: What's the chance of getting a 6 or a heads? Well, of course, one way to do it is to look at those 12 things that could happen. When you roll a die, it could come up 1 through 6, or a coin could come up heads or tails. We have those 12 possibilities. We can just look at the chart and half the time it is going to come up heads. Six of those 12 have a heads in them; heads with a 1, 2, 3, 4, 5, or 6, or it could come up six times with a tails.

There are seven possible ways in which you could get a 6 or a heads. There's only one way you could get a 6 and a heads—only one out of those 12. We have this distinction, then, between the "or" and the "and." Now often, it's much more difficult to figure out a probability of getting "this *or* this," whereas it's easy to think of getting "this *and* this," because if you want to get both a 5 and a heads, you have to multiply the probabilities. You have one-sixth of a chance of getting a 5. One of six times, you're going to get a 5 on this, and then, only among that one of six times will you be able to flip the coin and hope to win, and that will happen only half the time, and therefore, it is one-sixth times one-half. Whereas the "or" problem has a little bit of difficulty to it, because, if we're talking about 6 and heads, you could either get a heads or a 6, and they sort of overlap with each other.

Well, in that circumstance, it is often easier to look at the opposite question, so this is a case where we're going to take the view of looking at the opposite question. What is the probability of not getting a 6? Well, it is five out of six because you can get 1, 2, 3, 4 or 5. The probability of that, then, is five out of six. What is the probability of not getting a heads? One out of two. You can easily compute the probability—5 out of 12—by multiplying those two numbers together: $\frac{5}{6} \times \frac{1}{2} = \frac{5}{12}$. We've changed an "or" problem, which was a complicated one, into an "and" problem, which was an easier one, by stating the probability of avoiding the "6" multiplied by the probability of avoiding the "heads." We'll see another example of this in a minute.

Let's see another example of this, but it's an important principle of probability that turns out to be very useful, because it is often easier to find an "and" than it is an "or." I wanted to illustrate it with just one example. The only reason I am bringing this up is because it talks about the history of probability.

There was a French gambler by the name of Antoine Gombauld, the Chevalier de Méré—I have always wanted to be a chevalier; I don't know what that is, but it sounds great. I don't know how you become one, but I'd love to become one. Anyway, though, he was a gambler, and the reason that he's important is because he was playing two different gambling games, and he won the first game. Then, he tried to build another game that was very similar to it that he thought he would also win, and he started to lose. So he wrote a letter to Blaise Pascal, who was a famous mathematician, and said, "What's going on here?" Pascal wrote to Pierre de Fermat, and Fermat and Pascal, by answering this letter from this gambler, developed the basic principles of probability that we're talking about now—this question of "and" and "or," and the multiplying of probabilities, and so on. Thus, it has an interesting history.

I'll just take a minute to describe the two games that he had, and we won't talk about them too long, because they are sort of boring. Here was his first game. The first game was very simple. He took a die, just like this one, and he said, "I will roll the die four times. If I get a 6 in any one of those four rolls, then, I win. If I never get a 6, then I lose." Then, he would bet people even money, and he would win.

Let's think about this game, and we will very quickly analyze it. What is the chance of getting a 6 in four rolls? Well, that is sort of a hard question, because you could get a 6 the first time; or you could miss a 6 the first time and get it the second time; or you could miss the 6 both the first two times and get it the third time; or, you know, you could get it the second time and the fourth time. I mean, it is sort of a complicated thing to get a 6, or more than one 6. You see? It's complicated.

Think of the opposite question, though. This is the strategy of probability. You look at the opposite question, if you have a hard question. The opposite is that you would be missing a 6 every single time, you see? If the opposite of winning, of getting a 6 one of the first four times, is to miss the 6 the first time, roll it again, get something other than a 6, roll it again, get something other than a 6, roll it again, and get something other than a 6, four times in a row.

That is an "and" question, you see? You have to miss it the first time, you have to miss it the second time, you have to miss it the third time, and you have to miss it the fourth time. The chance of missing it the first time is very easy to compute. It is 5 over 6, right? Because there are five misses, and only one 6. The chance of missing the second time is 5 over 6. If you have to do both of them, you multiply them together, so that is 5 times 5, over 6 times 6, $\frac{5 \times 5}{6 \times 6}$. The third time is 5 over 6, the fourth time is 5 over 6, so that the probability of missing all four times is simply $\left(\frac{5}{6}\right)^4$. You can see on your screen that that value is 0.4822. What that means is that slightly less than half the time, you are going to miss 6 four times in a row.

Well, the opposite of that are the percentage of times that you are going to get at least one 6. Well, the opposite of that is then slightly better than 50 percent, right? In fact, it is exactly 0.517 and so on, so that is a little bit more.

Well, then, the Chevalier de Méré decided to change the game. He got bored with just one die. He decided to look at two dice, and he asked the question, "I'm going to throw two dice, and I am going to talk about getting two 6s, a pair of 6s." Of course, though, it wouldn't be fair to just do four times. That would be ridiculous; the chance of getting it four times wouldn't be right. Let's think of what would be the number of times you might expect to be the same game?

Well, here was his reasoning. He said, "One out of six times that I throw this first die, it is going to be a 6. Then, for each of those times, I want to shake four times. I want four opportunities to try to get a 6 with the other die. Then, I will throw both dice 24 times, six times four times, and then I will bet the same even money that I get a pair of 6s in those 24 times."

He played this game, and he lost. He started to lose, and he couldn't understand why he was losing, so he wrote this letter to Pascal, and Pascal wrote to Fermat. They got together and invented the concept of probability, because, you see, if you actually multiply it out—here it is; the chance of not getting a pair of 6s in a row is 35 out of 36, because, you see, the 36 possibilities for these two dice to come up—this could come up 1 through 6, this could come up 1 through 6, this could come up 1 through 6—there are 36 equally likely things that could happen, and only one of them is the 6 and 6. The other 35 are the bad ones, then, 35 out of 36. What's the chance that you roll the dice 24 times and fail to get two 6s any time? Well, it's $\left(\dfrac{35}{36}\right)^{24}$, okay?

Now, if you actually do that multiplication, you find that probability is 0.5085. That is slightly above half. You see? The opposite of that, then, which is what Gombaud was betting on, that is, getting a pair of 6s, is now less than half, 0.49, just a little bit less than half. So, he was losing just a bit. The probabilities differ, because you have to look at it by looking at the opposite.

That's enough about the history of gambling, because what I want to do is move forward from the ancient history to much more modern times, and in fact, I want to talk about a famous TV show that you have all either seen or heard about. The famous TV show is called *Let's Make a Deal*®. Do you remember the TV show *Let's Make a Deal*®? Some of you may remember; some of you may not. It depends on whether you had a misspent youth. *Let's Make a Deal*® was a TV show that had Monty Hall, the host, the emcee of the show, and the way it worked was that Monty Hall would come into the show and onstage, there would be three doors. Here, I have these three boxes that represent the three doors that were onstage.

He had some member of the studio audience come on down; he would talk to the member of the studio audience, and here is what he told the member of the studio audience, "Behind one of these doors is a fancy car, and if you take the right door, you will get this fancy car, you'll be able to drive home, and you will be rich and happy the rest of your life." That's what he said. Then he said, "Go ahead and choose a door. Choose any number you want." Then, the audience would erupt, "Choose One!" "Choose Two!" "Choose Three!" They were all yelling at each other. Finally, the guest would pick a number, and so the guest would pick Number, let's say, One. The guest would pick One.

Monty Hall did not open the door to Number One, though. That was not what he did. Instead, he said, "Okay, Number One. Well, I will tell you what I'm going to do. I'm going to open another door." Now, of course, Monty Hall knew which door the car was behind, so Monty Hall said, "I'm going to open another door, an alternative door," and there was a hush in the audience, you know, "What's there? I don't know. I don't know." But actually, there was no reason for a hush, because he never opened the door with the car behind it, because he knew where the car was, and he wouldn't do that. That would just cheat the guy out of the car, so he opened another door. Here he went; he opened this door, and what was behind this door? It was this thing. This is not a good prize, so let me just—this is to demonstrate something that you do not want. The guy was very happy; he said, "Whew! Oh, boy, I'm glad that it's not that, because now I know that the car's behind one of these two doors, either door Number One or door Number Two."

Then, Monty Hall said the following, "Okay. Would you like to stick to your original guess and take what's behind this door, or would you prefer to switch and take door Number Two?" You see, there were two doors that were closed, now, and either you could stick with your original guess, or you could switch. He would pose this opportunity to the guest, and then the whole studio audience would erupt in excitement: "Switch! Switch!" "Stick! Stick!" they would yell. Then the person would go back and forth, being swayed by the popular opinion, and didn't know what to do, and couldn't decide whether or not to stick or switch.

Now, this is something to think about. They were two doors. The car was behind one of them. It seems right off the bat that maybe it didn't matter. In other words, the probability was about the same that the car was behind one or the other. It didn't seem like so much. Well, let's pause before we make this decision. You might be on the show someday. It's worth while to pause and think about what you really should do at this point, because it turns out that this is an important question, and in fact, an important question in probability, because you would want the car. You see? You would want the car.

Here's how you think about it. Now, there are many different ways to think about this kind of a probability question. One way would be to simulate the situation by actually performing experiments. You know, you could have a friend of yours put cars behind different doors, and do the experiment over, and over, and over again, and just see how many times you've won the car if you switched, and how many times you've won the car if you stuck to your original guess. That would be an excellent way to get some intuition about whether it would be better to stick or to switch, or whether it is a 50-50 proposition. If you did it, maybe, a thousand times, you would get a sense of that. That's a good way to do it.

I advertised at the beginning of this lecture, though, that a good method of thinking about things is to exaggerate, because if you exaggerate a situation, you possibly have a different sense of what is going on.

Let's go ahead and try this, then, and we will exaggerate the situation in the following way: Suppose we go back to our original situation. The guest comes up, the guest chooses door Number One, and Monty Hall picks one, but instead of having just three doors, suppose now that Monty Hall had more doors, so that instead of just three, Monty Hall had six doors. I will just leave them like that. There were six doors, then. Now, you don't have as much chance of getting the car. Once again, the person chooses door Number One. Then, Monty Hall, instead of just opening one door, does the following. He says, "You know what? It's not behind door Number Three." Then he says, "You know what? It's also not behind door Number Four. It's not behind door Number Five—this is certainly not behind door Number Five, and it's not behind door Number Six."

You see? You don't want that. Nobody wants these things, incidentally. You don't want those things. Now, let's see if our intuition is being trained at all. Are we seeing anything here? If Monty Hall opened all of those doors—now, there are still just two doors that are unclear, and we don't know which one the car is behind. Do you still feel that there is an equal chance that the car is behind either of these two doors, or has your intuition about the situation altered?

Well you know, for some people, it changes a little bit. They get a little bit more puzzled by it, and some people think, "Well, it still sees 50-50."

Let's exaggerate some more. Okay? Let's think again. Now, this is just a mental experiment. We're doing a mental experiment. Suppose instead of having either three doors or six doors, suppose you had a hundred doors, and you were a member of the studio audience. You come into the studio, and Monty Hall said, "What door would you like?" And you picked your favorite number between 1 and 100. You picked 37. That's your favorite number, right? You picked 37, and Monty Hall said, "Well, I'm not going to tell you whether the car is behind door Number 37, but I will tell you what I will do. I'll open 98 other doors." He opened door after door after door after door; he opened all the doors except for door Number 62, and, of course, he left door Number 37 closed.

Now, how is your intuition doing? Do you want to switch, or do you want to stick? Now, Monty Hall, of course, knows where the car is. He would never open a door that had the car behind it, so, let's see, is your intuition right? Okay, let's exaggerate some more. Why stick to a hundred doors? Suppose we had a billion doors, okay? This is a real game show, so you are the contestant, there are a billion doors, and Monty Hall says, "Choose a door." You say, "Okay, I choose door Number 659,123,246." Monty Hall says, "Well, you know, you might be right. I'll tell you what I will do, though. I'll open 999,999,998 doors." He opens all of the doors except door Number 157,435,451, and your door. You'd be really impressed if I could say that number again, wouldn't you be? That's not going to happen.

Now, you see these two doors out of a billion. Now, you ask yourself, "Should I switch, or should I stay?" Well, probably at this point, your intuition is swaying in the direction of the switching side, because even though there are still two doors, and the car is behind one of those two, you are nevertheless pretty sure of where that car is, because how hopeful were you that you were going to drive away in a car when there were a billion doors, and you chose some number at random? You weren't very confident. In fact, you knew perfectly well that that was a ridiculous thing for you to even waste your afternoon doing, because you are not going to win that car, you see? We know how big a billion is. You are not going to win that car. After Monty Hall gives you the hint, though, of opening every single door except for one other door, and he knows where that car is, you are pretty confident, in fact, that by switching, you're going to win that car.

Let's take, now, our intuition; we've now crafted a new intuition about this, that in the case of a billion doors, it seems pretty clear that you should switch. If he opens all the doors except for one other, you should switch. You're really confident of that. Can we measure it? Can we be more precise, to say, "What's the chance that we are going to drive away with the car?" When there were a billion doors, you chose one, and then Monty Hall opened all the other doors except for one, and you switched. What's the chance that you were going to drive away with the car?

Let's think about it. Once again, let's think about the opposite. What is the chance that you got the car right on your first guess? Well, that's an easy question, right? That's an easy question. It's an easy question because there were a billion doors, and you had no idea where it was. You chose one, and your chance was one in one billion. That's your chance that you correctly guessed where the car was.

Now, suppose you were incorrect, which of course, you were almost certain to be, and you switched. What would happen? You would drive home in a car, because if you were wrong the first time, and he'd opened every other door except for one, if you would switch, you would get a car. Your chance of winning the car, then, if you were to switch is 999,999,999 out of one billion. That's your chance of coming home with the car if you were to switch.

Okay, now, we've got an intuition, and we've worked through some math to put some numbers to it, so we've quantified what our intuition led us to. Let's now revert back to the easier case that we first started with, and see whether we should stick or switch, and whether we can measure the probability of winning the car if we stick, and the winning the car if we switch. Okay, so let's go ahead and do this.

Now, we sweep away physically, and in our mind's eye are working, intuition-building analysis, and we come back to the original setting. Once again, you're the contestant, you walk up, you say, "I would like to take what's behind door Number One; that's my provisional choice." And Monty Hall says, "Well, look, door Number Three has this beautiful thing that you don't want." Now, this is great, and you say, "Okay, then, should I stick, or should I switch?"

Well, think about the analysis. The analysis in the case of a billion was: If I was wrong the first time, and I switch, I will win a car. The analysis is the same here: If I were wrong in my original guess, and the car is not behind door Number One, then, if I switch, I'm going to get the car.

What is the chance that you missed the first time? Well, that is easy to compute. The chance that you missed is two out of three, and therefore, if you switch, you have a two out of three chance of winning the car. Whereas, if you stick, you only have a one out of three chance of winning the car. Therefore, being wise people—and of course, this took most of the show, too, to go through this analysis, and you no doubt were talking it over with Monty Hall during the performance; I'm not sure if he would allow this kind of analysis, but it's very important—what you would do is you would switch, and sure enough, look at that. You win this beautiful, beautiful car that everybody wants, and to emphasize it, if you had made a poor probabilistic decision, you would have gotten this extremely ugly monkey. This is an analysis of a Monty Hall problem, and it has been sort of intriguing. It was in the Marilyn vos Savant column for many months, and a lot of people argued about it, because their intuition said, "Look, if you've got two doors, and you don't know which one it is behind, the chance is 50-50." Even, I'm embarrassed to say, some mathematicians did this. That was a bad thing. We don't want to talk about them.

Okay, now, I want to end with one sort of intriguing feature of probability that involves dice, but these dice are a little bit different. Let me show these dice, and what we're going to do is play a little game with these dice. It's an interesting game, and I always play it with my students, and when I play it with my students, I always bet them, because I like to win money, so you see, I bet them. With this game, though, it seems like a game where the other person is sure to be able to win, and yet, I will show you why that may not happen.

These are interesting dice, and let me show you these dice. They all are dice in the sense that they have six sides. They're all cubes, but they are unusual in that the spots on them are not the numbers 1 through 6. For example, do you see this die here? If you look at it carefully, every single side is a 3; every single one of the six sides is a 3. Isn't that funny? This is a silly die, because no matter what you roll, you get a 3. That's pretty nice. There you go.

Do you see this die? This one is interesting in that it just has two numbers to it. It has the number 5, and there are three sides that have the number 5. The other three sides have the number 1, see? These are funny dice, aren't they? They are a little bit odd. Here's another one. This one has 6s and 2s. It has two 6s. You see a 6 over here, and a 6 on the opposite side, and all of the other four sides are 2s. This one is also interesting. It's got four 4s. All these are 4s, but two of the sides don't have anything on them at all. They're blank; they are zeros.

Now, these are funny dice, and I will tell you the game. Here's the game. The person comes up, and they can choose any one of these dice that they want, and I choose another one. We roll the dice against each other, and whomever's number comes up that is better wins that round. We roll 20 times against each other, and whoever wins the most number of times wins the game. Okay, is that clear? You just pick two dice, and you roll them against each other.

You see, my student can come in and analyze these dice, and choose the best one. For example, if you look at this die with all of the 3s, this die is much better than the die that has two 6s and four 2s. Because whenever the 6 and 2 die comes up, four out of six times, it will come up a 2; and the die that is all 3s will win two out of three times. This die that is all 3s is looking pretty good. Then, in fact, if you look at this die that has three 5s and three 1s, look. This die with two 6s and four 2s—whenever this die comes up with a 1, the 6-2 die will beat it, and whenever this comes up with a 6, it will win even if you've got a 5 on this one, so the 6-2 die is better than the 5-1 die. In fact, it wins two-thirds of the time. A similar analysis shows, because you've got—this one has four 4s and two blanks—whenever you get a 5 on this die (5-1), it's going to win regardless of whether you get 4 or 0; if you get a 1 on this die (5-1), it will win if you get 0s on this die (4-0), so it wins two-thirds of the time.

It looks like the die with all 3s is the best one then, because it beats 6-2 die two-thirds of the time. 6-2 beats 5-1 two-thirds of the time. 5-1 beats 4-0 two-thirds of the time. Isn't that great? You should then choose the die that has all 3s. There's only one small wrinkle. This die, the die that has four 4s beats the die that has all 3s two-thirds of the time. Right? Because as long as you get a 4, you win. The dice are not arranged in a linear order, then. They're actually arranged in a circle, so that no matter what die my student chooses, I can choose one that will beat it two-thirds of the time. Now, this is the kind of dice game that we all want to play, and we can strive for in our lives to maximize our success, so thank you.

Lecture Twenty-One
Great Expectations—Weighing the Uncertain Future

Edward B. Burger, Ph.D.

Scope:

Probability and statistics enable us to better understand our world. They move us from a vague sense of disordered randomness to a sense of measured proportion. They are the mathematical foundations of common sense, wisdom, and good judgment. They weigh our expectations and give us a refined sense of valuing the unknown. Perhaps, too, they enable us to view our world more truly for what it is—a place where the wonderful and the desolate alike follow rules of the aggregate while leaving individuals to their wild variation and unbridled possibilities.

In this lecture, we will put a number to the possibilities of the unknowable future and give a quantitative measure of the expected value. We will show what value to place on decisions concerning investments, games of chance, and life insurance and see that paradoxical situations sometimes arise. The expected value of an outcome is an important piece of information to use in making critical life decisions.

Outline

I. In this lecture, we will learn to put a value on all possible outcomes of a given situation.

   A. This exercise brings up the concept of *expected value*, which we can think of as a weighted average of all possibilities.

   B. Mathematically, the expected value of an outcome is the sum of the value of each possible outcome multiplied by the likelihood that each outcome will occur.

   C. Roulette offers a good example. If you place a $1 bet on any number in roulette and that number comes up, you win $36.

       1. A roulette wheel has the numbers 1–36, as well as 0 and 00. Thus, any spin of the wheel has 38 possible outcomes that are equally likely.

2. The probability that any one number will appear is $\frac{1}{38}$. The probability of losing is the opposite, $\frac{37}{38}$.
3. Because you paid $1 to play, the profit you would receive from winning is $35. This is the value of winning.
4. We multiply 35 by the likelihood of winning, which is $\frac{1}{38}$.
5. We then add to that result the expected value of the alternative, which is losing. We know that the likelihood of losing is $\frac{37}{38}$. The value of losing is –1, because it costs $1 to play.
6. Adding those two numbers, we get $-\frac{2}{38}$, or $-\frac{1}{19}$, of $1, which is –5.3, or a little over 5¢. The negative tells us that, on average, we will lose money at the roulette wheel.
7. The 5.3¢ is, of course, a running average of what we will lose in playing roulette a number of times.

D. This game could be made more fair if the casino paid out $38 each time you won. Then, the expected value of playing would be 0, and if you played the game a number of times, you would break even. A *fair game* is any game for which the expected value is 0.
  1. An example of a fair game would be betting $1 on the flip of a coin and, if the coin lands as you called it, winning double the amount you bet.
  2. If you played this game a number of times, your losses would equal your winnings. In other words, the expected value of this game is 0.

E. This example leads us to a doubling strategy for gambling.
  1. With this strategy, you play the same coin-flipping game, but if you win, you quit, and you have made a $1 profit.

2. If you lose the first flip, the doubling strategy prompts you to play again and to double your first bet. If you win this second round, you still make a $1 profit.
3. If you lose the second round, you double your bet again, bringing your total investment in the game to $7. If you win in this round, you will receive $8, again netting a $1 profit.
4. This pattern continues; ultimately, when you win the coin toss, your profit will be $1, which is also the expected value of the game.
5. If we calculate the likelihood of each successive coin toss going a certain way, the result is an infinite line of fractions: $\frac{1}{2}, \frac{1}{4}, \frac{1}{8}, \frac{1}{16}$.... Adding these fractions, the result is 1.
6. This strategy is a sure thing, but you may have to lose a large sum of money before you win $1.

II. Another coin-tossing game that is a sure winner, at least from an expected value point of view, is called the *St. Petersburg paradox*.
   A. This problem was devised by the Swiss mathematician Daniel Bernoulli in 1738.
   B. You pay an undisclosed amount of money to play the game, which again, involves flipping a coin.
   C. If the coin comes up heads, you win, and the game is over. If the coin lands on tails, you get to flip again.
   D. The payoff for landing on heads with the first toss is $2. If you get tails on the first toss and have to flip again, you'll get $4 if you land on heads on the second toss. The payoff doubles each time you land on tails and have to flip again.
   E. What is the expected value of this game?
      1. The probability of tossing heads at the beginning of the game is $\frac{1}{2}$, and the payoff is $2; thus, the value of this outcome is $\frac{1}{2} \times 2 = 1$.

2. The probability of tossing tails on the first toss and heads on the second is $\frac{1}{4}$, and the payoff is $4. The value of this outcome is, again, 1. In fact, the value at every stage of the game is 1. Adding all those stages together, the expected value is infinite.

F. How much would you be willing to pay to play this game? If you paid $1,000, for instance, you would have to toss nine tails in a row to make back your investment. The probability of getting nine tails in a row is $\frac{1}{2^9}$, or $\frac{1}{512}$, or .00192, a very slim probability.

III. We often engage in forms of gambling in our everyday lives without even realizing it.

A. An insurance policy is a good example.

B. In decision making, we are also, in a sense, calculating the expected value of one outcome over another.

C. We can examine another paradox, proposed by William Newcomb, that involves some philosophical aspects of decision making.

1. In an experiment, a scientist leads you into a room containing two boxes. You can see that one box has $10,000 inside and the other is opaque; you have no idea what is inside it.
2. After the scientist leaves the room, you can choose to keep the contents of both boxes or only the opaque box. The opaque box may be empty, or it may contain $1 million.
3. Obviously, you would choose to keep the contents of both boxes. However, the experiment has another condition, which you are told about before you make your decision.
4. The condition is that if the scientist believes that you will choose both boxes, she will leave the opaque box empty. If she believes that you will take only the opaque box, she will put $1 million in it.
5. Suppose you also know that the scientist's predictions are accurate 90 percent of the time.

6. What is the expected value of taking both boxes? The probability that the opaque box is empty is .9, because the scientist is correct 90 percent of the time. The probability that the opaque box has $1 million in it is .1. If you compute the expected value, the result is $110,000.
7. What is the expected value of taking only the opaque box? The answer is $900,000.
8. This paradox raises a number of philosophical questions and shows that thinking about expected value helps quantify decision-making issues.

D. You might also ask whether you should spend $100 on lottery tickets or invest $100 in this course. If you calculate the value of buying lottery tickets, you'll find that for every $1 you spend, you lose about 50¢. Clearly, *The Joy of Thinking* is a better investment.

IV. In our lives, we face many issues that don't seem to be quantifiable, but we may be able to calculate an expected value or weigh likelihoods and payoffs to reach a qualitative expected value that will help us make more informed decisions.

Suggested Reading:

Edward B. Burger and Michael Starbird, *The Heart of Mathematics: An invitation to effective thinking*, Key College Publishing, Section 7.5, "Great Expectations: Weighing the unknown future through the notion of expected value."

Questions to Consider:

1. You wish to invest $1000 and you have two choices. One is a sure thing where you will make a 5 percent profit. The other is a riskier venture. If the venture pays off, you will make a 25 percent profit, otherwise you lose your $1000. What is the minimum required probability of this riskier venture paying off in order for the expected value to exceed the value of the first investment?

2. In Newcomb's paradox, first suppose that the psychologist just flips a coin to determine whether or not to place the million dollars in the box. What is the expected value of selecting both boxes? What is the expected value of selecting just the Zero-or-Million box? Suppose, instead, that the experiment was run by an all-knowing, god-like being. What would you do?

Lecture Twenty-One—Transcript
Great Expectations—Weighing the Uncertain Future

In the previous lectures that Mike has shared with us, we have been thinking about the idea of probability, that somehow, we can put a numerical value to things that have yet to occur. The unexpected, the future, we can somehow weigh. Today, what I would like us to do is to put an actual value on all possible outcomes, to see what value the future might actually hold in store.

Louis Pasteur once wrote: "Chance favors only the prepared mind," and when you think about this course as a course involving the joy of thinking, I can't think of anything more appropriate than thinking about the prepared mind, especially if we're going to take risks, and quite often, we take risks every time we walk into a casino; in fact, we absolutely do take a risk when we walk into a casino.

Today, then, what I want to do is take a look at how you can measure that, how you can be prepared for those unexpected losses, and this really brings up the value of *expected value*. That's the value of a particular event happening, and the way that you can actually think of expected value, in a sort of humanistic way, is to think about it as a weighted average of all possibilities.

Now, explicitly, mathematically, we define the idea of expected value to mean listing all of the possible outcomes that we can have, seeing the value of each of those outcomes, and taking each value of each outcome, and multiplying it by the likelihood that it occurs. Thus, if an outcome is actually very likely to happen, we actually weigh that very heavily. We multiply it by its probability, which would be very, very high. If another outcome is, in fact, very, very low in its likelihood to occur, then, we multiply it by its probability, which would make that value much smaller. The idea, then, is that we take the value of each possible outcome, and multiply each possible value by the probability that that outcome will occur, and then add all those numbers up. That would give us the expected value.

That sounds like sort of a complicated thing, but let's actually look at an example, and see that, in fact, we witness this all the time. The moment you step into a casino, you have actually lost, I'm sorry to say, but that's probably true. For example, consider roulette. Roulette is a really fun game. As you know, roulette has lots of numbers,

numbers from 1 to 36. You can pick your favorite one, the day that you were born. In my case, I would bet 9, which I always do, and rarely win, and you can bet whatever number you want. It turns out that if you place a $1 bet on a number, like, let's say, 9, if you win, you get $36. Thirty-six dollars—that's huge. Just a $1 bet, if you win, gets you $36.

Well, this sounds like a wonderful deal, and certainly one that you should consider taking—or is it? Well, let's compute the expected value. You can think of your birthday, or the month that you were born in, and these are numbers that sometimes reside between 1 and whatever, but we go up to 36. You get a $36 payoff, but it turns out that those aren't all the numbers, because in fact, as some of you may recall, there's also 0, and there's even 00. It turns out that 0 and 00, in fact, are how casinos make their bread-and-butter when it comes to roulette. Now, let's see why.

It turns out that since we have 0 and 00, there are a total of 38 equally likely possible outcomes, because we could spin a number from 1 to 36, but then we also have 0 and 00, so that's 38. What is the probability, then, that any one particular number appears? Well, that probability would be 1 out of the 38 equally likely possibilities, so we see 1 out of 38. That is the probability that we will actually get the number we bet on.

What is the probability of losing? Well, it is the opposite. Mike talked about that in Lecture Twenty. We see that it is actually going to be 37 over 38, because there are 37 numbers that are not our number out of 38 equally likely. It is therefore very, very likely that, with 37 over 38, we will lose. It will not be on our number.

Now, if we were actually to make the bet and win, they give us $36, which sounds wonderful, but you have to remember that you paid a dollar to play, so what is actually the profit that we get? The profit that we actually get is $35. In fact, then, the amount of money that we increase our wealth by if we win is only $35, so that that is the value of winning. It is actually $35. It is the input we get minus the amount that it cost us to play, and we have to consider that. In this case, then, it is $35.

We take that value, $35, and now multiply it by the likelihood of getting that outcome, which we already saw was 1 over 38. Now, we add to that the alternative, namely, losing. Well, if we lose, what is the value of that event? It cost us a dollar to play, and we lost, so we lost a dollar, so that the value is actually negative, because we lost, –$1, and what is the likelihood of that? Well, the likelihood is 37 over 38, so it is very likely that we will lose. If we add those two numbers up, what do we see? When we add those two numbers up, we get –2 over 38, which reduces to –1 over 19, of a dollar, and what is that? One-nineteenth of a dollar is just a little bit over 5 cents, so it is actually going to be 5.3.

Now remember, if you actually compute that, if you see the chart, it's negative. The reason that we see –1 over 19, or –5.3 cents is because it shows that, on average, we're going to be losing money. The negative is actually deficit spending. Let's roll, and see what happens. If you actually turned the roulette wheel, I will bet on 9, and you bet on your favorite number. Here it comes, and we're landing on—I can't see it—on 00. Wow. Well, that is, in fact, what they love, because they throw those numbers in there deliberately to make the odds of winning even less. I would have lost my dollar.

Now, I lost a dollar on that bet. I lost a dollar, but I claim that the expected value is, in fact, losing 5.3 cents. Well, it turns out that nobody loses 5.3 cents. Nobody does, because if you bet a dollar, you will lose a dollar, or you will win a $35 profit. Where is the 5.3-cent loss? Well, it turns out that that is a running average, what you expect to have happen if you keep doing this again, and again, and again, and again. On average, we're going to see that each person loses about five cents. Some will win, and a lot will lose. On average what you see, though, is that the casino makes five cents, and that is what the casino counts on. The casino wants to make five cents every time you play roulette, and that's how it makes its money. If they make enough five-cent profits, they've really done well.

How could you possibly make this game more fair? Well, you could actually make it more fair by having the casino actually pay out more money. For example, if the casino were to pay $37 for every time that you do this, then, if you think about it, what would happen in that case? In fact, let's actually say that they pay $38, so that they pay for as many times as there are numbers on the wheel. Well, then, it works out, if you compute the expected value, to be 0.

What does it mean for the expected value to be 0? You put a dollar down, and you get $38 back in this case. Well, in that case, what you would see is that if you played this game again and again, you should break even. In fact, then the casino would not be making money on average, and you wouldn't be making money, and we call that a *fair game*. A fair game, then, is any game where, in fact, the expected value works out to be 0, and there are many examples of a fair game.

Let me just show you one. Suppose that you take this, just a regular fair coin, and you flip it, and you bet, say, some amount of money, and if you win, you get double that, so that, in fact, if you win, you get that money back, plus the same amount that you put in. For example, if you were to bet a dollar, and you won when you flipped it. You said, "heads," and it came up heads, then you would win two dollars. That first dollar would cover how much it cost, and you would make a dollar profit.

Well, that would actually be a fair game, because the expected value would be 0, meaning that if you kept playing this long enough, and kept betting a dollar each time, you would expect, on average, to in fact, see no wins, and no losses. Your losses would equal, would match out, your wins.

There are some interesting strategies, though, that I want us to think about, some strategies that actually seem a little bit surprising, potentially. There is the doubling strategy. Let me show you the doubling strategy. The doubling strategy says, "We're going to flip the coin, and if it comes up, we win." We bet a dollar, and if it comes up that we win, then we stop. Notice that we get paid two dollars, and in that case, we have made a one-dollar profit, and then, we stop, even though you might be saying, "Oh, I'm on a roll. I'm going to do it again." The doubling strategy says, "Stop right there." Quit, and you have made a dollar profit.

Now, on the other hand, suppose that you lost. Suppose that you actually bet the dollar, and in fact, you flipped, and you didn't win. The doubling strategy says that you need to play again, and this time, you double your previous bet, so that if you bet a dollar before, you now will bet two dollars. Now, look what happens when you flip the coin this time. If you were to actually win, you would get twice what you just bet, which is two dollars, so you would get four now. Notice that the first three dollars cover your initial investment, and you still remain at profit of one dollar, so that you still make one dollar. You still make one dollar.

What if you were, in fact, to lose the second time? You've invested three dollars. The first one dollar bet, you doubled it the second time, you lost, and you now bet again, but what's the process? You double your previous bet, so now, I bet four dollars. Now, I have invested a total of seven dollars. What happens if I win now? If I win now, I get a total of eight dollars, because I just invested a bet of four dollars. The first seven dollars cover my losses, and look what happens. I get a dollar profit.

Well, this pattern actually continues. If we keep using this doubling strategy, when we finally do win, what we will see is that we will make that one dollar profit. We will get a lot of money, but the profit is only one dollar. Well, this sort of interesting, because if you think about it, you say, "Gee, this is a wonderful strategy. I am guaranteed to win a dollar. I just use the strategy. Namely, I keep playing and doubling my previous bet until I win, and once I win, I stop." That's the rule. You must stop, and then you have won a dollar.

Well, if we actually work out the expected value of this, let's see what we get. You can see that the expected value will always be a dollar. It will always be a dollar. Why is that? Because the first time through, the probability that the game would stop the first time—if the game stops the first time, that means that you have won the first time out, and if you have won the first time out, what's the probability of that? Well, it means the coin landed the way you wanted it to, and the probability of that is 1 over 2, one-half. With one-half the probability, then, you will see a payoff of these two dollars.

Okay, great. Now, we have that number, which is what? Well, it's one-half, and what is the profit? The profit is 1, so the value of it is 1, and the probability is 1/2, so we take 1 times 1/2 and get 1/2.

Now, we add to the probability that we didn't win the first round. What would happen then? Well, then, we bet two more dollars, and if we win the next round, what is the value of that? The value, again, remains one dollar. There is the one dollar profit, the value to us, and what is the probability of winning the second round? That means that we first flipped it the coin one way and it didn't come up, and we flipped it another time, and it went the way we wanted it to.

What is the probability, then, of getting those two coin tosses to be the right way? Well, there are four possible values for the coin toss that we saw earlier, in Mike's lecture. We could get "heads-heads," "heads-tails," "tails-heads," and "tails-tails." Four equally likely; one of those will produce this outcome, so it is one-fourth. One over four then times the value, which is 1.

Well, this pattern actually continues, and what we see is that the value is always 1, and what are the probabilities? The probabilities are always half of the previous probabilities, so 1/2, and then 1/4, and then 1/8, and then 1/16, and we just add up all those fractions.

Well, it is an infinite list of fractions that we have to add up. How can we actually add them up? Well, if you look at a number line, you can see what the answer is. Imagine that this is a number line, and if you look down, what you see is that this piece right here is a half, so there is the first half. Now, we add to that a quarter, so I just going to take this and add this next half, here—here's the whole number line—we add the next half, which is now a quarter, and if we take half of that, that is an eighth, and we add that on, and then a 16th, and we add that on, and a 32nd, and so forth, and we keep doing this, and what happens? It fills up this line segment perfectly, so that if you add up all those infinitely many fractions, what do we see? We see the number 1.

Well, that is not surprising. If we play this strategy, we would expect to win one dollar, right? Because you keep doing it, and you win one dollar, so that's right. This seems like a great way to win a dollar, where if you do this each time coming get $1 million, and you can do $1 million each time, instead of a dollar. What's the problem? The problem is that while it's a sure thing, we may have to lose arbitrary amounts of money before we actually win that one dollar. You might have to keep investing, and investing, and investing, and investing, so if you don't have an infinite amount of money, then, this game might not actually be the best game for you, because you have to be able to stand all those losses before you finally win, and then, what do you win? You win a one-dollar profit on a potentially billion dollar bet. Thus, it's interesting, but maybe not the best strategy.

Well, how about one where you're guaranteed, actually, to do really great, at least in the eyes of expected value? This is a wonderful paradox called the *St. Petersburg paradox*, and it was actually first discovered and reported by the Swiss mathematician Daniel Bernoulli, in 1738. It's a wonderful paradox, and let me tell you about the game.

The game is, again, to flip a coin, but the rules are a little bit different now. Now, I'm not going to tell you how much money it costs to play the game. I am going to leave that to the end. You're going to pay some amount of money yet to be disclosed, and then, we're going to tell you what happens next.

What happens next is that you're going to flip the coin. If the coin comes up heads, congratulations, you have won, and you win two dollars, and the game is over, and that's the end of the game. Not bad. Okay, fine.

Now, if you flip the coin, and it comes up tails, that means that you get to flip again, so, in fact, you're just going to keep flipping as long as you see tails, and the moment you see heads, the game will end. The first time that you flip a heads, you are going to see, in fact, that the game ends.

Now, what is the payoff? If you see heads the first time, you get two dollars. If you first flip and see a tails, and then you flip and see a heads, then, you get double that; you get, actually, four dollars.

Suppose that, in fact, you flip "tails-tails." Then, you flip again. If you see heads now, you get double that; you get eight dollars.

This is looking pretty good now, right? You want, actually, to flip lots of tails in a row, because that's going to build your bank account. This was, then, flipping once, and then, if you flipped again, which he sees you get, one, two, three, four, five, six—oh, my goodness; so there, you get four dollars, and now, if you were to get "tails-tails" and then get heads, then, in fact, you get eight more dollars—one, two, three, four, five, six—it's getting up there, so it's going off the table now. You can see, then, that the money is really growing. It's exponential growth. In fact, Mike talked about this at the beginning of Lecture Two.

The point is that every stage you are going to get double what you had at the previous stage. Well, this looks like a great investment, and, in fact, if you compute the expected value, what is the expected value for us? Well, let's see.

At the very beginning, the probability of flipping heads is actually one-half, so it is one-half, and what is the payoff? The payoff is, in fact, two dollars, so it's 2, times the probability, which is one over two. Notice that the 2 upstairs and the 2 downstairs cancel, so the expected value of that one thing is 1.

Now, the other possibility is that you got tails, and then got heads. What is that? Well, there, you get four dollars. Four dollars is given to you, and then what happens? Well, what is the probability of that? Well, the probability of that is a quarter, because you first have to get tails and then heads. One-half times one-half is one-fourth, and then you multiply that by the value, which is 4. Again, then, we see 4 divided by 4. We see 1, again, and that pattern continues. At every stage, when we take the probability, which will be 1/2 to some power, and we multiply it by the value, which is going to be 2 to that same power, those terms cancel out, and I get 1. I can potentially play this game, then, forever, and when you add up infinitely many 1s, what do we see about the expected value? The expected value, in fact, turns out to be infinity. 1 plus 1 plus 1 plus 1 plus 1, etc.

This is a game, then, where the payoff, if you do it again and again, is going to be infinite. It is going to be infinite, so this is actually a fantastic game to play, you know? Where do I sign up for this game?

Well, the answer is: How much are you willing to pay at the beginning of this game to play? Well, it sounds as though, if the payoff is infinite, I should be willing to pay just about anything. It turns out, though, that this is a great paradox. Suppose that you spent merely $1000, merely $1000 to play this game whose expected value is infinity. How would you actually make money on that $1000 bet?

Well, you would have to get a long run of tails, because if you got heads too early, you wouldn't recoup your $1000 investment. You only need a little bit. You're either going to get two or four, or eight, and so forth. How many do you actually have to get? You actually have to get nine tails in a row in order to make back that $1000 investment. What's the probability of getting nine tails in a row? It is 1 over 2^9, which is 1 over 512, which works out to be 0.00192. That is so slim, and it seems so counterintuitive, that if you just bet that $1000, you'd expect to win infinitely many dollars, but it turns out that the probability that you will win is very slim.

What is the moral here, then? The moral here is that you win very, very, very seldom, but when you win, you win big. It is a huge payoff, but the probability of winning is very, very low, and so those things balance somehow, in some sort of complicated way. You're seeing a wonderful paradox where the expected value is infinite, and yet, if you put up $1000, the chances of you making money is very small. This is a wonderful paradox known as the St. Petersburg paradox.

It turns out that it's not just gambling in casinos where we experience these things. In fact, quite often, we engage in many forms of gambling in our everyday lives, and don't even realize it. Even if you don't like casinos, and are offended by them, it turns out that we are gambling. For example, consider the notion of insurance. If you take a look at an insurance policy, in fact, the insurance company actually wants you to live a long, long time, and keep paying those premiums. In some sense, we all want to die early, to make the big payoff. Now, of course, we don't really appreciate the big payoff, but our heirs do, and so, again, what is the expected value of that?

Even in decision-making, this becomes an issue. Even when we make decisions as to whether we want to do one thing, or the alternative, we are actually computing, in some sense, an expected value, putting a quantitative measure to it.

It turns out that there's this wonderful paradox. In fact, the French philosopher William Newcomb proposed this paradox, and it actually remains unsolved today. I wanted to share it with you, because, in fact, it's a terrific paradox, and it involves an experiment. We are led into a room, and there's a woman who is going to give us this experiment. She's a scientist, and she leads us into the room, and in the room, we see two boxes. One box is completely opaque, and we have no idea what is inside. The other box, we can actually see, and enjoy, that there is, in fact, $10,000 in cash in that box.

This is exciting. Now, the scientist tells you that she's going to leave the room, and when she leaves the room, you'll have two choices. You can look at these two boxes and either select both boxes, and keep the contents of both boxes, or you might prefer just to select the question-mark box, leave the $10,000 behind, and only take the contents of the question-mark box.

Well, if that is your decision, then there are a couple things that you will wonder about. First of all, what is in the question-mark box? Well, I'll tell you. The question-mark box is either empty, so that there's nothing in the question-mark box, or the question-mark box is filled with $1 million. The box, then, is either empty, or has $1 million in it. Let's review the possibilities again. You can either elect to take both boxes and their contents, or you can leave the $10,000 behind, just take this box, which may contain $1 million. Well, it seems pretty plain what you would do, right? You would take them both, because then you have a chance of getting $1 million, and you're definitely guaranteed to get the $10,000. It sounds like a no-brainer.

There are a couple of added wrinkles here, though, you are told about before you make your decision. Namely, how is it determined whether the million dollars is put in this box, or not? Well, this scientist is an excellent judge of the human psyche, and so she sizes you up, and she will make a prediction as to what she believes you will do when you are put in that situation, and if she believes that you will leave this $10,000 box and just take the question-mark box, then, she puts $1 million in it. If she believes, though, that you might actually try to be a little bit greedy, and take them both, then, she puts nothing in this box.

Beforehand, then, she makes a decision. She meets you, and shakes your hand, and talks to you, and if she believes that you're the kind of person who actually will take them both, then she puts nothing in here. If she believes that you are the kind of person who will actually leave this, and just take this, then, she puts $1 million in there.

Well, this raises the question: How good is she at predicting the human psyche? Suppose she's very good, and suppose that we know, that we're told beforehand, that she's 90 percent correct. That means that 90 percent of the time, and she accurately predicts a person taking both, and s/he does, or she accurately predicts a person just taking the one, and s/he does 90 percent of the time.

The question is, now, that you are led into the room, and you are faced with these two boxes. You can't lift them up, by the way. Some people say, "Well, I'm going to see if it's heavy." You can't touch them. The question is: What would you do?

Well, this is a wonderful paradox, because there are several ways of looking at it. One way to look at it is to say, "Well, let's compute the expected value." What is the expected value, then, of taking both boxes? We are tempted to take both boxes, because she has already put the $1 million in there, or not.

With probability of 0.9, you're only going to get $10,000, but with the probability of 0.1 that she's wrong, you get $1 million and $10,000. When you work out that expected value, then, what do you see? You see an expected value of $110,000, so that on average, you would expect to get $110,000 from that strategy. That seems great.

What if you just took the $1 million box, and left this one aside? Well, with 0.9 probability, very likely, she guessed that, and therefore, she put $1 million in there, so that you would get $1 million, with a probability of 0.9. That is really high. With a probability of 0.1, however, she actually got it wrong. She put nothing in there, and in fact, you didn't take this, so you got nothing. Well, when you work out that probability, you see that you get $900,000. Just by taking the question-mark box, the expected value is much higher, almost nine times higher. You get $900,000 on average if you just pick this versus taking them both, when on average, you would only get $110,000 with this.

What do you do, then? Well, the paradox is, the truth is, that it's a done deal. Once she has either put the money in or not, it's done. The question is, then, do you feel special? Do you feel like you're one of the people that could slip under that 90 percent radar, and actually have her make a mistake, so that you take them both, in the hope that she has, in fact, predicted that you would only take this one, and put $1 million in there? Or do you believe in free will? Can you come in there and just say, "You know what? I'm going to live my life not as a sheep, but I'm going to take them both, and that's it." Free will, and whatever happens, happens, and it's a done deal, and it's finished. Or do you say, "I'm just one of many, many people, and I'm ordinary, I'm average, because the truth is, almost all of us are ordinary," in which case, in that case, she would predict us correctly, and we should just take this box.

The question is, then, do you feel special? Do you believe in free will? It all boils down to this amazing paradox, which, in fact, has no answer at this point, because if you work out the expected value answer, it's clear that you just take this. If you think about free will, and the fact that, "You know what? I don't care," you can take them both.

It is a really fascinating question, and the interesting thing about the question is that you can vary it by asking, "What if there was only just a dollar here?" Well, then, certainly, I think almost everyone would take this, because it's like buying a one-dollar lottery ticket. You potentially lose a dollar, but with 0.9 probability, you win $1 million. In fact, then, there's a wonderful paradox that shows us that thinking about expected value allows us to actually quantify the issue.

We see this in our everyday lives. For example, should you buy $100 worth of lottery tickets, or should you spend $100, let's say, on this course, *The Joy of Thinking*? Well, you can actually compute the value of this, right? In fact, spending $100 on lottery tickets, on average—if you average a lot, most people lose, but of course, when you win, you win big. It is very much like some of the puzzles and paradoxes in the gambling games that we saw earlier this lecture. It turns out that on average, for every dollar that you invest, you get about 50 cents back, so you lose half your money. For example, then, the expected value of $100 worth of lottery tickets would be about -$50. You would lose $50 on that investment, whereas if you were actually to watch the lectures of Mike and myself, and somehow it informs and inspires you to think and look at things differently, and think more effectively, then, in fact, you might actually be getting salary raises, making better investments, both in your time and money, and, in fact, I think it's clear where you should put your $100 for investment.

Anyway, you can see that life decisions really do, quite often, boil down to looking at the quantitative aspects of things, and, you know, I think it really highlights the notion of weighing all possibilities. Quite often, then, when we walk through our lives, there are issues that arise that don't appear to be of a quantitative nature. We are making decisions about how we're going to live our lives, and where we're going to live our lives, what investments we're going to make, and not just investments in terms of financial issues, but also investments in terms of our time, our resources, and allocation issues.

Quite often, by computing the expected value, either if you can actually calculate it knowing the likelihood of various outcomes and their values, you can just take those various products, the value of an outcome times its likelihood, and add all those numbers up, or you can just do it in a qualitative fashion, and think about how likely something is, and the potential payoff. Does that outweigh the other alternatives, together with their likelihoods and their payoffs?

When you put these ideas together, all of a sudden, you see your journey through life in a more directed way. You see that there are alternatives. One of the greatest lessons, in fact, that economics teaches us is the notion of opportunity cost. Right? Every decision we make comes at the price of all the other alternatives, and the notion of expected value allows us to actually quantify all those things, and then, when we make a decision, we can really view it as being an informed decision, a wise decision, at least knowing the fact that we are not certain of the outcome. When things lie ahead, and our future is in front of us, and we don't know for certain what will happen, quite often, weighing all the possible likelihoods is a powerful way of proceeding.

The thinking lessons from this journey into the notion of expected value, then, really are several-fold. First of all, there's the idea of looking at all possibilities, examining the issues from several points of view, as we did, in fact, with the paradox of Newcomb, where if we look at them one way, we want to take them both, and it's the done deal, or the other way, saying, "I'm not special. I'm just like everyone else. I will pick the $1 million question-mark box with the hope that the woman is correct, and get $1 million."

It is also the notion, though, of looking at things not only from several points of view, but looking at them with a quantitative eye, for the moment that we start to associate numbers, in this case, likelihoods, with our lives, all of a sudden, we begin to make far more informed decisions.

The next lecture will take up the notion of what it means when things, in fact, are completely unpredictable, and we will take a look at randomness, and we will see that even within randomness, we're going to find structure. In the next lecture, then, we will take look at randomness, and then, we will close with the ever popular notion of coincidences.

Lecture Twenty-Two
Random Thoughts—Randomness in Our World

Edward B. Burger, Ph.D.

Scope:

Coincidences and random behavior do occur and often with predictable frequency. A bit of careful thought reveals that coincidences are not as shocking as they may first appear. One of the most famous illustrations of randomness is the scenario of monkeys randomly typing Hamlet—*Hamlet happens*. Other examples, such as *Buffon's needle*, show how random behavior can be used to estimate numbers, such as π. The theory of random walks is filled with counterintuitive and surprising outcomes and appears in *Brownian motion* and the stock market.

We can apply the principles of probability to understand coincidences and random behavior. The basic definition of probability and how to compute it allow us to gauge more meaningfully how rare or common a seemingly unlikely event really is. We can estimate the probability that no coincidences will occur. This opposite view helps us to understand how likely coincidences really are.

Outline

I. These last two lectures before we close the course discuss the ideas of randomness and chance.

    A. Even in the world of randomness and chance, order exists.

    B. This lecture illustrates the thinking strategy of understanding the issue. Sometimes, the first step in solving a complicated problem is to accurately define and grasp the issue at the heart of the problem.

II. We begin with an incident that occurred on June 3, 2003. On that date, Sammy Sosa sheared his bat in half as he hit a ground ball to second base in the first inning of a game against Tampa Bay. It turned out that he was using a bat that was hollowed out and filled with cork.

    A. Was his use of this bat intentional? Officials confiscated the 76 bats in Sosa's locker and found that none of them had cork inside.

B. Some argued that the fact that none of the other bats were filled with cork proved that Sosa's choosing the one bat was unintentional.
   C. We know that the chance that Sosa chose that bat at random from a group of 77 bats is $\frac{1}{77}$, a very slight probability.
   D. We'll return to this incident at the end of the lecture and look at it in a different way.
III. Precise and quantitative information can flow naturally from random behavior or processes. A classic illustration of this is the experiment known as *Buffon's needle*.
   A. In this experiment, we randomly drop needles onto a sheet of lined paper to see how many needles will cross the lines of the paper when they fall. What is the probability that a needle will cross a line?
   B. This question was first raised and, later, answered by the 18th-century French scientist Georges Louis Leclerc, Comte de Buffon.
   C. The surprising answer is that the probability that a needle will cross a line can be computed and is exactly equal to $\frac{2}{\pi}$. Remember that π is the circumference of a circle that has a diameter of length 1.
   D. Using the random data gained from dropping needles onto a sheet of paper, then counting the number of needles that cross a line, we can actually estimate the precise number π.
      1. To do so, divide the total number of needle drops by the number of needles that cross a line and double the result. The answer will be a close approximation of π.
      2. It seems remarkable that a random process of dropping needles will lead to π. Indeed, the more needles that are dropped, the closer the estimate will be to π.
   E. Buffon performed this experiment by tossing breadsticks over his shoulder onto a tiled floor. Hundreds of years later, atomic scientists discovered that a similar needle-dropping model seems to accurately predict the chances that a neutron produced by the fission of an atomic nucleus would either be

stopped or deflected by another nucleus near it. Even at the subatomic level, nature appears to drop needles.

IV. A *random walk* is another idea that comes up in our exploration of randomness.

- **A.** Let's examine a situation in which a pedestrian takes one step either to the right or left, depending on the outcome of a coin toss.
- **B.** Under these conditions, what are the chances that the pedestrian will return to his or her starting point?
- **C.** This question seems enormously complicated, but let's try to understand the issue.
    1. For the pedestrian to return to the starting point, every move to the right must be counterbalanced by a move to the left.
    2. This knowledge tells us that to return to the starting point, the pedestrian must flip the coin an even number of times.
    3. We also know that, using a fair coin, at some point in the walk, the number of heads will equal the number of tails, which means that the pedestrian will return to the starting point.
    4. We know with certainty, then, that at some point, the pedestrian will return to the starting position.
    5. We can expand the question by asking: What is the likelihood that the pedestrian will reach a point five steps to the left of the starting point, or –5 on a number line? To reach this point, the pedestrian would have to flip five more tails than heads in the process of flipping the coin.
    6. In fact, with an infinite number of coin tosses, the pedestrian will reach every point on the path or number line.
- **D.** What happens if we add another dimension to the pedestrian's path? We flip two coins; one of them tells the pedestrian to move right or left, and the other, to move up or down.
    1. We can think of this question in the same way, but the mathematics is a little more complicated.

2. In this situation, too, the pedestrian will reach every possible point in the plane with probabilistic certainty.

E. What would happen if the pedestrian could fly and move in three dimensions? Although it is by no means obvious, it is not certain that the pedestrian will return to the starting point. This situation offers too many degrees of freedom, and in fact, the probability that the pedestrian will return to the starting point is .699....

V. The classic question of randomness is: Would an infinite number of monkeys typing randomly eventually produce *Hamlet*?

A. This issue was first raised by Sir Arthur Eddington, the famous astronomer, in 1929.

B. Suppose that we boil down all of *Hamlet* to just the number 3. If we roll a six-sided die enough times, will we eventually roll a 3? Certainly.

C. Now suppose that *Hamlet* was 1 on a yellow die and 2 on a green die. As we saw earlier, the probability of rolling this combination is $\frac{1}{36}$. The probability is slimmer than it was with one die, but again, if we roll the dice enough times, we will eventually roll this combination.

D. Now imagine a die with 48 sides, one for each character and symbol on a keyboard. Although it is very slim, the possibility exists that such a die would land with the appropriate sequence of characters to produce the precise wording of *Hamlet*. If we roll the die infinitely many times, then with probabilistic certainty, that sequence will eventually happen.

E. With a million monkeys hitting the typewriter keys at a rate of one key per second, we would have to wait 10^{60} years for one of them to type: "To be or not to be. That is the question."

VI. In this lecture, we have seen random processes become very precise. Indeed, the idea of the random walk has applications in science in *Brownian motion* and in stock market modeling. We have also learned that even though the probability of an event may seem slim, such as Sammy Sosa's picking the cork bat at random, given enough random trials, the event will occur.

Suggested Reading:

Edward B. Burger and Michael Starbird, *The Heart of Mathematics: An invitation to effective thinking*, Key College Publishing, Section 7.3, "Random Thoughts: Are coincidences as truly amazing as they first appear?"

Questions to Consider:

1. Using a piece of graph paper, a penny, and a dime, embark upon a random walk on the grid. Flip both coins. If the penny lands heads up, you move one unit to the right, if it lands tails up, you move one unit to the left. If the dime is heads, you move one unit up and if it is tails you move one unit down. Take 50 steps and mark your trail on the graph paper.

2. Suppose we have a monkey typing on a word processor that has only 26 keys (no spaces, numbers, punctuation, etc.) The monkey types randomly for a long, long time. What is more likely to be seen: MICHAELSTARBIRD or EDWARDBURGER? Explain your answer.

Lecture Twenty-Two—Transcript
Random Thoughts—Randomness in Our World

Well, in the last lecture, we saw how we could actually assign a value to an event that has yet to occur, in particular, the idea of the expected value of a particular event, and have see how it, in fact, informs potential decisions, even looking at gambling in a different way.

These last two mathematical lectures before Mike and I close the course will revolve around the idea of randomness, and then the notion of coincidences. These are things that we experience in our everyday lives, and it turns out that through a very interesting careful way of looking at things, through a mathematical mindset, if you will, we can actually bring some understanding and some type of analysis, to see these things in greater focus.

Really, then, although each individual outcome of a random process can't be determined, nevertheless, the aggregate, if we look at it all, in fact, leads us to predict that certain things are inevitable. Really, then, in some sense, from randomness comes order, as we'll see.

Bertrand Russell once wrote: "How dare we speak of the laws of chance? Is not chance the antithesis of all law?"

Well, Bertrand Russell was certainly a very wise and sage philosopher, but the answer to this question is, in fact, that law does exist, even in the world and realm of chance, even within the realm of randomness.

Now, the real life lesson for this particular lecture is the fundamental and very basic idea of understanding the issue, the idea that sometimes, we're faced with a problem, a circumstance, that is very complicated. The first step to really cracking that open is to actually understand what the issue is, and we'll see reflections of that very basic life lesson, if you will, several times this particular lecture.

Let me begin by recalling something that you may remember. Way back in June of 2003, while playing Tampa Bay, Sammy Sosa sheared his bat in half as he hit a ground ball to second base, I think, in the first inning. It turned out that that bat was hollowed out and filled with cork. Well, this led to a huge thing. Was it intentional, or not? The officials confiscated all the other 76 bats that were, in fact, in Sammy Sosa's locker, and they checked them all out, and none of them had cork. So that people were saying, "Oh, see, it was obviously unintentional," and people were saying, "Oh, it was intentional," and so forth.

Well, here are some basic questions. Suppose that he were to actually pick out his bats at random. He's about to go up and hit the ball, and he randomly goes into the now-77 stash of bats; 76 of them are perfectly fine, and that last one, in fact, is cork. What are the chances that if he picked it out at random, in fact, he pulled out the cork bat?

Well, we know the answer to that. We've been studying probability. The answer is that there's one way of succeeding; only one cork bat out of the 77 equally likely—if he's picking them out at random—possibilities, so 1 over 77, which is a very small number. In fact, then, when people were saying, "Gee, it was obviously unintentional," maybe that's not so clear, because if he were just to pick one out at random, the chances of him getting that particular bat, the cork bat, are very, very low. That then might give some credence to the idea that, in fact, it was deliberate.

I want to return to this example at the very end of the lecture, to see if we can look at it in a completely different way, after we have explored these ideas of randomness.

It turns out that precise and quantitative information can actually flow naturally from random behavior, or even random processes, and I want to begin with an interesting and classic illustration known as *Buffon's needle*. This is a really neat experiment that you can actually try if you want to. It just involves taking a collection of parallel lines, as you can see here, and a bunch of needles, which I will depict here using a red needle, and the needles need to be the exact same length as the width between the lines, so the can see that it fits right in there.

Here is an interesting probability question, then: Suppose that we randomly dropped these needles—and I'm going to try to emulate that here, but you have to understand that when I say "randomly dropped," it even includes the orientation, how they are being pivoted, and so forth. I randomly drop them, and you will notice that there are several possibilities. I dropped three. You'll notice that sometimes, the needle crosses one of the lines, and sometimes, for example, right here, the needle doesn't. What we see here, then, that is that, in fact, in this very small sample size, two out of the three times, the needle crosses the line. The question is, then: What is the likelihood that if we did this again, and again, and again—could we figure out the precise likelihood that, in fact, the needle will cross the line?

Well, this is known as Buffon's needle, and this question was actually first raised, and then immediately answered, by the 18th-century French scientist George Louis Leclerc Comte de Buffon. Now, his name is so long that if you dropped his name on the piece of paper, you are guaranteed, of course, to hit a line, so that, of course, is clear.

Well, the surprising answer is that the probability that the needle we dropped will hit a line can be computed exactly, and that exact number turns out to be 2 divided by π. Now, maybe you will remember π from your early mathematical school days, or career. The number π, in fact, represents the circumference or the length of the circle whose diameter has length 1. The number π, then, always was involved in finding areas of circles or lengths around, also known as circumferences of circles, and so forth. Thus, π is a very, very important number that we have all seen at some point in our careers, in school or otherwise, and the number actually works out to be 3.14159, and it keeps going, forever, in fact, and that special number turns out to appear in this probability. The probability that we actually hit a line turns out to be 2 over the special number π.

Well, in fact, mathematicians actually are very excited by the prospect of understanding π, in particular, computing the decimal digits of π to great accuracy. Well, this random event allows us to do just that, for, you see, if we were just to drop lots and lots of needles here, so that we would dropped a whole bunch of them down, and then count how many needles we dropped, and then compare that number to the number of those needles that actually crossed a line, we actually can get an estimate using that random data to the precise number of π, and here's how you would do it. It is actually quite simple. You take the total number of times you have dropped needles, and divide that quantity by the total number of needles that actually crossed the lines. Take that answer, and double it. That will give you an incredible approximation to π.

It seems sort of remarkable that a random event, of just dropping these things, will lead to a very precise notion of the decimal digits for π, and the more you do it, the more accuracy you get. In fact, if you go on to websites, if you actually have a tendency to surf the Internet, if you actually search for Buffon's needle, you can find simulators that actually simulate this, and you can see the accuracy building. As you drop more and more needles, this number gets closer and closer, in fact, to π.

Now, in fact, when Buffon—who actually did this experiment; believe not, he actually did this with French breadsticks, and I'm not kidding. You might think I'm making a joke, but I'm absolutely not, and what he did was he threw them behind himself in this restaurant, because the tiled floor turned out to have the exact same length as the length of the breadsticks, so he would just literally toss them back, and then he would count, you see, how many times he saw them cross, or not. We've all been taught not to play with our food, but here we see an example where this kind of random event really leads to something quite precise.

Okay, this sounds sort of frivolous, but really, it turns out that it's not. You know, hundreds of years after Buffon tossed his breadsticks, atomic scientists discovered that a similar needle-dropping model seems, in fact, to accurately predict the chances that a neutron produced by fission, of an atomic nucleus, would either be stopped or deflected by another nucleus near it. Even at the subatomic level, then, nature seems to be dropping needles. It also, I think, illustrates a wonderful life lesson, that sometimes, an experiment that may seem sort of frivolous and devoid of any real consequence or application, might actually be reflections of some greater idea, in this case, looking at, sort of, the physics of particles.

Anyway, that is Buffon's needle, and I hope that you got a kick out of that. In fact, though, when you are thinking about randomness, one of the things that people think about is the notion of a random walk. Now, a random walk, in fact, is just a regular walk that we can take, but where the direction that we move at any particular moment is completely determined by randomness, not by our doing, and you may say, "Well, gee, that's pretty easy. I can take a random walk. I will walk to an intersection of a street, and I will just sort of look up in the air, pick a direction, and go in that direction, get to the next intersection of streets, and then go up and say, 'Oh, I will just go that way.'"

That is actually not random, because we are making some decision, even if we think in our heads that it is random. It may be that we're saying, "Well, gee, the coffee shop is that way, so just to be random, I will go that way." Well, that now, is actually a very calculated decision. Actually, walking in a random fashion is quite difficult. How can you do it?

Well, you can let chance decide, and so, here, for example, let's simulate the notion of a random walk along a street. Here's our pedestrian, who volunteered for this experiment, and here I have a fair coin. What I'm going to do is flip the coin, and if the coin lands heads, then our participant will take one step to the right, and if in fact I flip and I see tails, then, in fact, she will move one step to the left. Okay? This is actually going to generate a random walk, so now, let me actually do this for real. This is live. You can't see it. That's heads, so she moves one unit over. Heads again, one unit over. Tails, one unit back. Tails, one unit back.

Actually, now, you will notice something sort of familiar. She looks around and realizes that this is exactly where she started, and this actually raises an interesting question. If you're actually embarked on a random walk along a street, what are the chances that you will actually, indeed, return to your starting position?

Well, it turns out that this question seems phenomenally difficult, because the whole thing is random. We don't know the actual particular moves, because every time we flip the coin we have potential getting either heads or tails, so we have to move forwards or backwards, and so it seems like it's completely irresolvable. It also seems like the answer may be that you might never come back. This might have been a real fluke, because we might have gotten a whole bunch of heads, to move us out here, and then a few tails, and then a whole bunch more heads, so that we just keep going further and further away, never actually coming back.

Well, how can you think about this? Here is a wonderful place where we can now take in, to internalize, the idea of the life lesson: "Understand the issue." Let's then move from this sort of chaotic thinking of "Oh, it's randomness; there's no hope of us understanding anything," to the notion of: What does it mean for this woman to return home?

Well, let's think about it. Even though we don't know the flips, we do know that as she moves. If she's going to return home, there's one basic element that we can conclude. For every time that she moves to the right, she must counterbalance that with a move to the left. Now, that actually is a major breakthrough in this analysis. Think about it. Even she takes a lot of steps to the right, she has to take that many more steps to get her back home. Now, they may not be in a row. For example, here's a scenario where we see two steps to the right, one step to the left, but now, we still owe ourselves one step to the left, and maybe, now, she keeps going to the right for awhile. Finally, in fact, I have to get all way back. Every single right move has to be countered by a left move in order to get back.

Well, that basic, fundamental principle allows us to see a whole bunch of things. First of all, if there's any hope of her getting back, she has to flip the coin an even number of times. For if she flips an odd number of times, that at the very best, she will be off either by 1 or by -1, or she might be far away, of course, possibly. We need, then, and even number of tosses in order to have any hope of this. Otherwise, she definitely cannot be back.

That's the first observation. Now, if this is a fair coin, that means that if we keep flipping it again, and again, and again, we would expect to see tails, on average, just as often as we would see heads. What does that mean? It means that if you keep flipping this long list of "heads, tails, tails, tails, heads, tails, tails, tails, tails, heads," and so forth, and you do that long enough, we would expect, and it would be pretty likely, that some point somewhere there will have just as many tails as there were heads. The moment we come to a point where the number of tails equals the number of heads, the matter how they are configured in that long list, we know that if she were to walk that path, she would end up back home.

This very complicated question, then, about flipping, and the chaos of walking back and forth, and the chaos of the coin turns out really just to boil down to the basics question of: If you have a fair coin, where it is equally likely to come up heads or tails, if you keep flipping it, again and again and again, what is the likelihood that at some point, there will be just as many heads flipped as tails? Well, intuitively, the answer is clear. The answer is: with certainty. At some point, I would expect to see just as many heads as tails, if I wait long enough, and that turns out to be the answer. It is with probabilistic certainty that she will always return back to her starting position, so that is an amazing insight, which, it turns out, just comes from understanding the issue. What does it mean to return back home?

In fact, you can actually take this a step further, and look at the following little extra corollary or consequence of this observation, which is: Suppose you actually wanted to have her hit the point -5. If she starts at 0, what is the probability, or the likelihood, that by just traveling in a random walk, in a random way, she will at some point hit -5?

Well, what does that mean? Well, if you use the same kind of analysis, what it would mean is that instead of actually having an even number of heads and tails, I have to, at the end of the day, if we keep doing this for a long time, at some point have exactly five more tails than I had heads. For in that case, even though I might be jiggling back and forth, I have one, two, three, four, five extra in this direction, which would bring me to here. What is the likelihood, then, if you are flipping a fair coin, that at some point, there will be five more tails than heads?

Well, the answer is that if you're going to be flipping it forever, you'd expect to see every configuration, including even this type of configuration. Therefore, what we conclude is that she's going to hit every point. Every point. If we keep doing this long enough, she will hit every single point on here with probabilistic certainty. A random walk, then—while it is completely random and we don't know the outcome, we can still say very precise things. We know, for example, that she will always return home, and in fact, if we let her continue to walk even after she returns home, she will, if you let her walk long enough—her poor legs would probably die out on us, but if she kept going forever, like the little bunny that we see on TV with the battery, what we see is that she will hit every point.

Now, what about if she were to come off this street and move on to other streets, in particular, move into the second dimension? In fact, then, we flip two coins, and one coin tells us whether to move up or down one unit, and the other tells us to go right or left. Well, you can actually think about it in the same kind of way, although the mathematics get a little more complicated. What you will see, though, if you try it and work out the mathematics, is that there, too, every possible point that she can hit in the plane, she will hit with probabilistic certainty, just by this random act of flipping two coins.

Here is the great surprise, though. Suppose that she was in three dimensions. Suppose that, now, she can start to fly, so that she doesn't only go up here and here, but she can now fly out into space, so that we use all three dimensions. One coin tells her which unit to move in this direction, the second coin tells her which unit to move in this direction, and the third coin tells her which unit to move in the "up" or "down" direction, such action starts to move out in points in space. It turns out, in this is by no means obvious, that it is not certain that she comes back home. In fact, as we saw in the lectures on the fourth dimension and dimension, there are too many degrees of freedom here. It is possible for her to actually get lost in space.

Now, you can actually compute the probability that she inevitably comes back home, and the probability turns out to be 0.699 something. In fact, then, the probability that she comes home is more than 50 percent, so it's fairly likely. It is almost 0.7, so it's almost 3 out of 4 chances of her coming back home, but is not certain anymore, because of the extra space. If you go to higher dimensions, if you go for a walk in dimension four, then it is even less likely to return home. As you make the space bigger, in fact, you have this issue of getting back home. Again, we see randomness giving rise to very precise information.

Well, there's a wonderful, classic example of randomness that I am sure you have heard of, and any lecture on randomness would be remiss without actually discussing it, and this is a question about monkeys typing at the typewriter, so the question is: Suppose you had monkeys typing randomly at the typewriter, completely randomly, hitting the keys at random, not knowing what they mean. If we have enough of them, and let them go long enough, would they eventually produce *Hamlet*?

Well, this very, very famous question, in fact, is one that goes back to Sir Arthur Eddington. In fact, he made this observation. Sir Arthur Eddington was the famous astronomer. He made this observation back in 1929, and he was trying to describe some features of the second law of thermodynamics, and he wrote the following: "If I let my fingers wander idly over the keys of a typewriter it *might* happen that my screed made an intelligible sentence. If an army of monkeys were strumming on typewriters they *might* write all the books in the British Museum. The chance of their doing so is decidedly favorable…"

Well, in fact, he has the correct answer. In particular, if we have enough monkeys typing randomly at keyboards, what we will see, in fact, is a perfect version of *Hamlet*. Now, that seems so incredible. How could randomness actually lead to such a famous book? By the way, I'm sure this is not how William Shakespeare wrote it, in case you're wondering. No, no, I think he actually was very deliberate—not random at all.

Actually, we can illustrate this, and make it more intuitive by considering a simpler issue. Let's actually boil down *Hamlet* to just one number. Let's suppose that *Hamlet* was just one number. Instead of lots and lots of words, and lots of footnotes, and lots of characters, and so forth, and scenes, let's just suppose that *Hamlet* was the number 3, and suppose that I had a die. This is now my keyboard, and I have a die, and I start rolling it. The question is, if I do this long enough, will I see a 3? Well, the answer is "certainly," assuming this die is a fair die, so that all sides are as equally likely to come up as any other. We know that, in fact, the answer is "yes."

Now, how do I know that? I'm going to prove it to you in a couple of different ways. First of all, just think of every time you have either rolled a die, or watched someone else roll a die. At some point in your life, you saw a 3, so that proves it, right? You all saw a 3. Additionally, though, you can just do it right now, live. Here, I rolled a 4; and there, I rolled a 3. In fact, then, if you roll this thing long enough, since each is as equally likely to appear as any other number, we see the probability is one out of six. It is therefore not very likely, but it certainly can happen, and if I do it with enough frequency, we would expect that to happen.

If *Hamlet*, then, were just the number 3, then it is very intuitive that if we kept rolling enough, or let the monkeys randomly hit the 6 keys, we would expect at some point that some monkey would hit 3, or, in our case, to roll a 3, and then, we would have *Hamlet*.

Great. From a mathematical point of view, once we have this idea, the whole issue is resolved, because all we have to do now is to make it a little bit more complicated, so instead of looking at *Hamlet* being just one number, let's suppose that, in fact, *Hamlet* is two numbers. Let's suppose that *Hamlet* is "yellow 1," and "green 2." That turns out to be *Hamlet*. That is a little more complicated. In fact, how many possibilities are there if I were to roll these two dice? Well, actually, Mike talked about this in Lecture Twenty, and he saw that there are six possibilities for this, and six possibilities for this, so if you look at all possibilities, you see that there are 6 times 6, or 36 total equally likely possibilities. The likelihood, then, of getting a yellow 1 and then a green 2 turns out to be 1 out of 36, so far less likely than just rolling a 3 on one die.

However, 1 out of 36, if I were to wait long enough, and keep doing this, what would you guess? Well, you would guess that if you keep rolling these dice enough times—oh, look at that, 3-3; that's not *Hamlet*—you would expect that maybe that's *Julius Caesar* or something. I don't know, but that's not *Hamlet*. We have to keep rolling. If the probability is 1 out of 36, though, which is slim but still possible, if we repeat that enough times, we would expect to see a 1 and a 2. Do you think, in your own life experience, watching people roll dice, you will remember that you once did see a 1 and a 2? That person produced *Hamlet*, if *Hamlet* were only two numbers.

Well, again, to the mathematician, that answers the question, because now, what do we have here? Now, we just have a die, if you will, with lots and lots of numbers on it, in particular, one side of the die for every letter, every character, and a space, and a period, and so forth, so imagine that really big die with, maybe, 48 symbols on it, since there are 48 symbols on a standard keyboard. Imagine a 48-sided die, if you will, and then, you start rolling it, and writing down what you see. Every letter corresponds to what we write down.

Really, then, *Hamlet* is just a really long, long, long, long sequence of dice rolls, and it is extremely unlikely that you're going to roll it billions and billions of times, and produce the precise sequence of rolls that spell out every single letter, every single space, in *Hamlet*, but that possibility, in fact, exists. It's not 0. It's very slim, and if you have a very slim experiment, in terms of the chances that it occurs, the chances are very slim that if you repeat that long enough, you will expect to see it. Just ask, if you repeat this long enough, you will see a 3. It is the exact same idea.

In fact, then, if you let the monkeys type away, what we do, in fact, see at some point, is *Hamlet*. Well, the thing is that sometimes, you might actually have to wait quite a while to see *Hamlet*, right? How long would you have to wait? Well, it turns out that you can actually use some calculations. I wanted to share one calculation with you. Suppose you took just a million, a mere million monkeys, okay? Now, of course, here, in my mind, when I do the experiments, I have infinitely many monkeys, and they are typing forever. Let's just take a million monkeys, and let's suppose that the monkeys are hitting the keys at a rate of one keystroke per second, so that's pretty fast.

Now, by the way, when we do this experiment, in our minds and in Sir Arthur Eddington's mind, we genuinely mean that the keys are being hit at random. If real monkeys try this—in fact, they tried this in England, believe it or not sometime back. They actually did an experiment, and it turns out that the monkeys might have been enamored with the with the "S." They hit the "S" a lot. They liked that key. That's not random. When we mean monkeys typing at the typewriter, we mean that every key is just as likely to occur as any other key is.

Suppose they are typing at one key per second, and we have a million of them going. Well, there are a lot of possibilities. How long might we have to wait in order to see someone type out, "To be or not to be. That is the question"? How long would we have to wait? Probably more than 10^{60} years. That's 1, followed by 60 zeros. That's how many years we would have to wait in order to expect to see just the phrase, "To be or not to be. That is the question." That seems like it's pretty slim, and it is, but if you wait that long, we would expect to see that phrase. If we wait even longer, we would expect to see more, and if we wait forever, we would expect to see, in fact, all of *Hamlet*.

It reminds me, actually, of a wonderful old routine that I believe is credited to Steve Allen, whom you may remember was the very first host of the Tonight Show, long before Jack Paar or Johnny Carson and Jay Leno. The routine was basically something like this. You see good old Steve Allen, and he's pretending to be a reporter. He says, "Okay, well, scientists have said that if we have enough monkeys type long enough, they will finally bash out *Hamlet*. Well, actually, scientists are trying that experiment right now, and let's see how they're making out." Then, there's somebody dressed up in a monkey suit hitting the typewriter, he pulls out the piece of paper, and reads, "To be, or not to be. That is the zymfanblot. So close," and then he tosses it back, and continues, "Back to the studio."

The idea is that somehow, you can get very, very close to *Hamlet* and still not perfectly be *Hamlet*. It also raises an interesting question, and the question is, not only will these monkeys bash out *Hamlet*, but every possible variation of *Hamlet*, and not only will you see every possible variation of *Hamlet*, but they're going to actually produce every single book; if you let them go long enough, they'll produce every single book in English ever written or that ever will be written, because every single book that's out there is just a collection of keystrokes in a particular order. However, as we are discovering here, we might have to wait a very long time to actually produce all of *Hamlet*, and we might have to, in fact, be satisfied with just having the monkeys type out and only produce *Macbeth*. Maybe you would take that anyway; I don't know.

Anyway, here we are seeing lots of illustrations where a random process actually leads to something very precise. In the case of the random walk, in fact, it turns out that the random walks occur in nature. The path, for example, of a liquid or gas molecule is determined by its knocking around and bumping into nearby molecules. This random path, then, and this notion of a random walk, turns out to be the idea behind what's called *Brownian motion*. In fact, the notion of randomness, whether it's random walks, or randomly typing, turns out to have implications even into the stock market. People actually try to model the behavior of the stock market through the idea of a random walk, and through this idea of randomness.

What's the fundamental idea, then, here? The fundamental idea is that sometimes, we actually experience randomness that seems completely unpredictable, and in fact, it really is unpredictable, but somehow, more importantly, if we produce a random event enough times, then, in fact, we can say something about that random event, something for certain.

Now, I want to return, finally, to the Sammy Sosa issue. Now, remember that if he's pulling the bat out at random, the probability that he picks the cork bat is going to be 1 out of 77, very unlikely. However, what if, in fact, Sammy Sosa's strategy, before he went out to bat, is that every time he goes to bat, he just randomly picks one of these bats? He doesn't look; he just randomly picks one, and goes out and hits. Well, now, what would you expect? Well, now, even though the chances are 1 in 77 that he picks that bat—very, very unlikely—if kept repeating that process, what would you imagine? At some point, he would, in fact, pick that cork bat, so maybe, in fact, it really was unintentional, that maybe, if he's picking at random, you expect that at some point, even an extremely rare event would occur.

The realization that something is possible, even if it's very, very unlikely, but given enough times will happen, really could inform our decisions, and not only that, but our behavior, and it allows us to look at our world and our actions in a whole new light. You know, one could actually view the monkeys pounding on the typewriter as an amazing coincidence. And you can think of a coincidence that way. In fact, in the very next lecture, what I want us to do its focus on the coincidences that we see in our everyday lives, and bring some structured, mathematical thinking to that. All of this came from the basic notion of understanding the issue: really, really focus.

Lecture Twenty-Three
How Surprising Are Surprising Coincidences?

Edward B. Burger, Ph.D.

Scope:

During the great Sammy Sosa-Mark McGuire home run race of 1998, Mark McGuire tied a home run record of 61 home runs on his own father's 61st birthday. What an amazing coincidence!

Coincidences are striking because any particular one is extremely improbable. However, what is even more improbable is that no coincidence will occur. We will see that finding 2 people having the same birthday in a room of 45 is extremely likely, by chance alone, even though the probability that any particular 2 people will have the same birthday is extremely low. If you were one of a pair of people in the room with the same birthday as someone else, you would feel that a surprising coincidence had occurred—as, indeed, it had. But almost certainly, some pair of people in the room would experience that coincidence. Coincidences and random happenings easily befuddle our intuition.

Outline

I. In the last lecture, we saw that the laws of probability enabled us to gain a better understanding of randomness.
  A. In this lecture, we'll learn the importance of experimentation as a thinking strategy in honing our intuition.
  B. Further, we revisit the central idea of focusing our viewpoint from a qualitative one to a quantitative one.

II. One aspect of coincidences is that we notice them.
  A. Most people are familiar with the famous list of Kennedy-Lincoln coincidences.
  B. Any one of these coincidences is surprising and unexpected, but we must ask: Could we find similar lists of parallels between any two presidents?
    1. Imagine the many thousands of potential parallels that were investigated to arrive at this list.
    2. We could examine this same list of potential parallels to identify many more differences than similarities.

3. In fact, if we examine the "life data" of any two people, we will find amazing similarities.
C. Another example of these parallels can be found in twin studies, which are significant sources of information on the "nature versus nurture" question. Nature versus nurture refers to the question of the extent to which personalities of people are determined by genetics versus upbringing.
   1. Twins who were separated at birth and brought up separately are excellent subjects for this nature versus nurture issue.
   2. Such a situation occurred in the 1970s with a set of twins who were separated at birth, then reunited when they were in their 40s. The similarities between the two were astonishing.
   3. Again, we should keep in mind that we will find amazing coincidences between any two people. To determine how amazing these coincidences are in a twin study, we must know how many questions the twins were asked to arrive at their list of similarities and whether the number of parallels between the twins is significantly greater than that between any two people.
D. If you enjoy coincidences, try the following experiment with two decks of cards.
   1. Shuffle each deck separately, then flip the cards over one at a time, simultaneously, from each deck.
   2. Will the exact same card appear in the exact same location in both decks? It seems unlikely that this coincidence would occur.
   3. If you try this experiment yourself, you will find that two-thirds of the time, you will flip over two of the exact same cards at the same time.

III. We are shocked by coincidences that occur in the public arena because they seem so unlikely and, often, eerie. For example, the New York Pick 3 numbers on September 11, 2002, were 9-1-1.
   A. Personal coincidences seem to occur more frequently; sometimes, however, we may have an experience that seems coincidental but is not.

B. Suppose, for example, that you receive an unsolicited e-mail from an investment company predicting that IBM stock will rise in the next week, and the following week, the stock goes up.

C. Over the course of the next nine weeks, you continue to receive e-mails accurately predicting the performance of IBM stock.

D. The following week, you get an e-mail stating that if you wish to receive the next prediction, the fee is $1000, which will be completely refunded if the prediction is wrong.

E. This offer seems to be very attractive, but it may be a scam that was structured as follows:

1. The company's first step is to make a list of 1024 investors and send them an e-mail. For 512 of the investors, the e-mail predicts, "IBM stock will go up next week." For the other 512, the e-mail predicts, "IBM stock will go down next week."
2. The following week, 512 e-mails are sent to the group that received the correct prediction. For 256 of these investors, the e-mail predicts, "IBM stock will go up next week." For the other 256, the e-mail predicts, "IBM stock will go down next week." At the end of that week, 256 people will begin to pay attention to the e-mails.
3. In succeeding weeks, the pattern is repeated. After nine weeks, two people will have received e-mails accurately predicting the future nine times in a row.
4. Finally, these two people are asked to send in a check for $1000 to receive the next prediction. One person receives an e-mail saying that the stock will go up, and one receives an e-mail saying that the stock will go down. The person who plans the scheme is guaranteed to be right in one case.

F. Thousands of people predict the stock market, and as we would expect, some of them are correct sometimes. We would also expect some correct predictions from the results of thousands of random attempts to predict the stock market by flipping a coin or throwing a dart.

IV. Another famous coincidence is the birthday question: What is the probability that 2 of 45 people attending a party will have the identical birthday?

A. Eliminating leap years, a year has 365 days. If there were 366 people at the party, we know from the Pigeonhole principle (discussed in Lecture Two) that they all cannot have different birthdays. With 366 people, then, 2 people at the party are guaranteed to have the same birthday.

B. You might argue that if we reduce the number of people to 180, roughly half of 366, then the probability would be $\frac{1}{2}$. If we reduce that by half, to 90 people, the probability would be $\frac{1}{4}$, and if we reduce the number of people to 45, the probability would be $\frac{1}{8}$.

C. This intuitive method of thinking about the question may seem reasonable, but it is not accurate. The reality is that with 45 people in a room, the probability that 2 of them share the same birthday is 95 percent. This coincidence seems counterintuitive, but we can retrain our intuition about this question by looking at it mathematically.

1. To understand this surprising probability, let's look at the opposite question: What is the probability that there are no birthday matches among these 45 people? That probability is actually very small, which means that the opposite probability is very large.

2. To compute the probability of not having a common birth date, we line up the people and start with the first 2. The probability of their having different birthdays is approximately $\frac{364}{365}$, or .9972, or 99.7 percent.

3. The third person must avoid the birth dates of the first and second people; thus, the probability that the third person has a different birthday is $\frac{363}{365}$, or .9945, or 99.5 percent.

4. As we continue down the line, the probability that we will not find a match decreases slightly, but it is still fairly likely that we will not have any birthday matches.
5. In Lecture Twenty, we learned that the way to find the probability that there is no match among all 45 people is to multiply the probability of each person's not matching; therefore, we multiply $.997 \times .994 \times .... 882 \times .879$, the product of which is approximately 0.05.
6. The probability that there is no match, then, is 5 percent; thus, the probability that at least one match exists is 95 percent.

V. Coincidences are more common than we usually believe they should be by chance alone. We need to retrain our intuition to understand randomness more accurately.

Suggested Reading:

Edward B. Burger and Michael Starbird, *The Heart of Mathematics: An invitation to effective thinking*, Key College Publishing, Sections 7.1, "Chance Surprises: Some scenarios involving chance that confound our intuition"; 7.2, "Predicting the Future in an Uncertain World: How to measure uncertainty through the idea of probability"; and 7.3, "Random Thoughts: Are coincidences as truly amazing as they first appear?"

Questions to Consider:

1. List three coincidences that you have experienced in your life.
2. Find two presidents of the United States who shared the same birthday. Are there two vice presidents who shared the same birthday? Are there two Democratic United States senators who have the same birthday? Republican senators?

Lecture Twenty-Three—Transcript
How Surprising Are Surprising Coincidences?

In the last lecture, we experienced the reality that randomness is not nearly as chaotic as we first thought. You know, really, sometimes people may think that randomness, by its definition, is that which can't be tamed, predicted, or understood. We saw in the last lecture, though, that by using a template of thinking, we were able not only to crack open the idea of randomness, but gain insight into it, and thus tame it.

In this last mathematical lecture before Mike and I bring the course to a close, we really want to take a look at the ever-present idea of coincidences in our everyday lives, or in the lives of our friends and loved ones. These events, these wonderfully surprising events, are just that, surprising, for the simple reason we are just shocked by a coincidence. The question is: Should we be? When you think about when we're surprised by anything in life, why is that? Well, obviously our intuition, our internal sense of how the world operates, must be running in some sense counter to the reality and thus, the surprise. Something happens, and we're surprised, because we expected something else. Therefore, what should we do in life?

What we should do is retrain ourselves, retrain our intuition, and build up a new suite of intuitions to allow us to see and perceive the world as it more accurately is. The question really is, then: How surprising are surprising coincidences?

The fifth century poet-philosopher-politician Boethius once said: "Chance, too, which seems to rush along with slack reins, is bridled and governed by law."

This quote, to me, really captures the spirit of what probability is, and we saw reflections of this, in fact, in our discussion on randomness. That, in fact, even within the random, we saw properties of law, things that we could say with great precision and perfect accuracy. And here we're going to apply those laws to the ever-present world of coincidences and try to tame them.

Now, the life lessons here are also very clear to me. First of all, when we find something that seems surprising, we need to explore and experiment in order to make those surprising ideas actually intuitive, so the notion of doing things, and trying, and experimentation, is important, and whenever possible, to move from the qualitative, fuzzy thinking that is so easy for us to sort of embrace, into a quantitative way of thinking, where we actually try to, as best as possible, in instances, put precision and numerical accuracy into our thoughts.

Well, one thing about coincidences is that we notice them, and I wanted to turn our attention to what are probably the most famous series of coincidences that people talk about, and they involve the famous Lincoln-Kennedy assassination coincidences. Here are some facts, then, that seem sort of eerie and chilling in their similarities. Abraham Lincoln was elected to Congress in 1846, while John F. Kennedy was elected to Congress in 1946. Abraham Lincoln was elected president in 1860, while John F. Kennedy was elected president in 1960. Lincoln's secretary was named Kennedy. Kennedy's secretary was named Lincoln. Andrew Johnson, who succeeded Lincoln, was born in 1808; while Lyndon Johnson, who succeeded Kennedy, was born in 1908. John Wilkes Booth, who assassinated Lincoln, was born in 1839; while Lee Harvey Oswald, who assassinated Kennedy, was born in 1939, and the coincidences continue. For example, John Wilkes Booth was known by his three names. Lee Harvey Oswald was known by his three names. Kennedy was assassinated with a gunshot from a book warehouse, and then the assassin hid in a theater, and vice versa with Lincoln. The list goes on, and on, and on.

The question is: Is there some sort of cosmic nature that is in the background here, that is putting these two moments in history, and these people and their lives, together, locked? Or, in fact, are we just noticing noise?

Well, one way to try to realize, or come to grips with the fact of whether something special is happening or not, is to ask: Are these parallels any more dramatic than any parallels that we would pick if we picked out two presidents at random, and looked at their facts? In fact, think of all of the facts for any one particular president, or, for that matter, for any one person, even yourselves. We come equipped with a long list of data: our parents, our grandparents, our children, aunts, uncles, cousins, best friends, where we went to school, where we grew up, the names of our pets, etc.

Now, you would have to make a very, very long list to encapsulate all the data. Famous people, of course, have more data. You know for someone who is secretary of state: Where he or she actually practiced law, where he or she actually represented people when they were in Congress. Now, if you made a long list of all these facts for Lincoln and for Kennedy and compared them, what would you see? You would see lots of differences. Lots of differences. One was from Illinois; one was from Massachusetts, and so on. Their wives' names are different. Their children's names are different.

You can make a list of the differences between these two presidents that is even more startling than the one of their similarities. If we have a long, long list, though, what would we expect, just from our random discussion before? We would expect that some things, in fact, would be parallel. We would expect those similarities, and in fact, if you pick any two people out, and look at those people's life data, we will find amazing similarities.

If we now, then, rewrite our questionnaire to throw away all the tons and tons of questions where, in fact, the answers were different, what we would see, in fact, is a surprising list of similarities. Both people were born in this particular month, or both people had wives that were born on the same exact numerical day. Every person's name, and their many acquaintances, then, and their dates, are all potential places for parallel.

The point is that we cannot know how this unexpected number of parallels are going to play out, but the truth is that, if we have a long enough list, we would expect those parallels to be there, especially between presidents, where, in fact, lots of famous data are known.

Another wonderful example of this is in the avenue of the scientific study of twin studies. Now, twin studies are significant sources of information concerning the "nature versus nurture" issue. Nature versus nurture, of course, refers to the question of the extent to which people's personalities are determined by genetics versus upbringing. Well, in fact, lots of experiments have been done, where we have two twins, a set of twins, that, for some reason, were separated at birth and raised completely independently, and then are brought together and reunited. The question is: How many similarities do we see?

In fact, this was done several times. The one case that I'm aware of happened back in the 1970s, and these were two infants that were separated at birth, never knew each other, grew up in completely different parts of the United States, and that were reunited when they were in their 40s. The similarities were unbelievable.

They both chain-smoked Camels. They both loved Miller Lite. They both actually had a winter house in Florida within three blocks of each other. They both, in their own regular homes, had big, big oak trees, one of which was surrounded by one of those white metal chairs that go around the entire tree. They both had that. They both had dogs named Skip. They both were married to wives named Mary, and then divorced, and then remarried to wives named Barbara. I'm making that up, but they had the same name. They each had two sons. The sons' names were the same, in each case. The list goes on, and on, and on. It's eerie, but the question is: Is this a nature versus nurture issue? The names of their wives: Is that somehow related to who they are?

Well, the answer is, if you pick two people at random, and start to ask lots and lots of questions, again, we will see those similarities. With these twin studies, then, we need to know how many questions were asked, and are the coincidences we are experiencing any more than we would expect to see with just two random people, who will, of course, enjoy many, many coincidences?

Thus, even with issues involving twins, then, it's not quite clear where the coincidences are genuine, or just there because of the random nature of coincidences.

Now, if you enjoy coincidences, then, you really have to try this experiment. Let me show you how you can actually potentially produce a coincidence if you want to. You need two decks of playing cards, 52 regular-sized playing cards, and you want to shuffle each deck up, so if you have a friend with you—in which case you should then definitely do this, because this could be potentially impressive. It also makes a wonderful little wager opportunity if you are so inclined, but if you are repulsed by that, then just do it as an experiment.

I have two different sets of playing cards, and I have shuffled each of them up. The question is: Have we just witnessed an amazing coincidence? Well, what would be a coincidence in this context? A coincidence would be, for example, if the exact same card would appear in the exact same location in both decks, namely, the same number, the same suit, in the exact same spot. Well, it seems very unlikely. There are 52 cards here, and here. They have all been shuffled up, and so forth. This seems like it would not be very likely at all. You should try it, though. Again, to re-hone intuition, you try an experiment like this.

Here, these do not match, and the question is: Will we see a match? Now, those are both fives, and they are both reds, but one is a diamond, and one is a heart. That doesn't count. I want a perfect match. The question is: Would we experience such a coincidence? Well, you can try this, and certainly, it doesn't seem very likely at all. In fact we would expect—our guess would be that we would not see a coincidence, and certainly, we're not seeing any now. That's wonderful. That is exactly confronting what our intuition tells us.

Actually, though, I urge you to try this experiment at home. Actually, genuinely do this for yourselves, and I think you will be surprised by the outcome. Because if you try this, again, and again, and again, you will discover that two-thirds of the time, more than 66 percent of the time, you will have actually experienced an amazing coincidence, that, in fact, there will be two cards that are identical at exactly the same location, you will flip over at exactly the same time.

There's a surprising reality, then, and our intuition has to conform to that, and we have to appreciate that, in fact, this intuition that we originally had was not accurate in terms of the understanding of the actual events.

Now, we are shocked by coincidences because they seem so unbelievably unlikely, and sometimes eerie, such as on the one-year anniversary of the September 11 terrorist attacks, when New York State picked three lottery numbers that were ominously 9-1-1. Now, personal coincidences occur much more frequently, and when we experience them, we can't help but tell our friends, our neighbors, and basically anyone who will listen, because they're so fantastic, but sometimes, maybe we are experiencing things that seem coincidental and, in fact, are not.

Thus, now I would like actually to offer you the following scenario that may have actually already happened to you, even in a different guise. Suppose one day, from the fictitious T. Rowe Smith Investment Company, you receive the following unsolicited e-mail. The e-mail says that next week, IBM stock is going to up, and they said to feel free to use this information any way you want to. Well, of course, the way you would use it would be to ignore it, because it was unsolicited, some spam, and so you probably would just delete that e-mail from the long list in your inbox.

Okay, but maybe you notice that, in fact, the following week, the reality is that IBM was up, so that was sort of amazing, and then you receive another e-mail from good old T. Rowe Smith, and here, the report, this coming week, IBM will be up. Okay, you ignore it again. Sure enough, though, IBM is up.

The next e-mail comes in the following week, and the e-mail reports that IBM will be dropping the following week. Sure enough, IBM goes down. Well, now you're taking note, because three times in a row, T. Rowe Smith was able to predict the actual, correct situation with respect to this particular stock. The next week, you get another e-mail, and the e-mail says, "IBM will go up." Sure enough, it goes up. You get another e-mail: "IBM will go down." IBM goes down. Another e-mail: "IBM will go down." The next week, it goes down.

It confirms week, after week, after week. Nine weeks go by. That's over two months, where you've been getting this data, and they have been right every single time. Well, you can't help but take that in and say, "That is an amazing coincidence," or maybe they really have some insight into IBM.

Then, in the next week, you get an e-mail that says, "We hope that you have enjoyed and invested wisely, and actually have been able to make lots of wise investments and grow your portfolio through our free advice. However, for the next week, if you would like our advice, we now ask that you please remit $1000. $1000, and we absolutely guarantee—we guarantee—that our results will be right, in the sense that if we predict incorrectly for the following week, we will refund your $1000."

Well, what do you have to lose? Interesting possibility. Interesting scenario. How would you react? Well, after all that data, where they matched it perfectly, right on the nose, $1000 seems like a very small price to invest to potentially make a huge windfall, if you hadn't already, by listening to them, and in the case that they're wrong, you would even get that $1000 back. It sounds like a great deal.

Now, what's really going on behind the scenes? Well, let me tell you what they could have done. They first take a list of e-mail addresses from people, and they take 1024 of them. They take half that list,

512, and tell them with an e-mail message that IBM will go up, and they take the other 512, and send the same kind of e-mail saying that IBM will go down. Well, what's going to happen? After the first week, there are going to be 512 people for which they inaccurately predicted the reality, so they just get thrown out. Those people get crossed off the list, but there are 512 people for whom they predicted correctly.

Then, that list gets cut in half. Now, they take 256 people, and tell them IBM will go up, and the other 256 are told IBM will go down. Will happen after that week? Well, there are 256 people for whom they were right twice in a row. Now, they repeat the process. They take the 256, and cut that list in half, and then cut it in half again, and in half again, and in half again, depending on how stock goes, up or down, until they are left with just two people, two people, and they are sent the e-mail message saying, "Send us $1000." Each of those people sends in $1000. They tell one IBM will go up, and the other IBM will go down. What happens? They know they will be wrong for one of them, and they send back that $1000 that they paid in, still holding the $1000 from the person for whom they guessed correctly, because they took out all possibilities.

From this vantage point, then, you see that the whole thing is a scam, and in fact, is illegal. It is almost like a pyramid scheme, in a way. The reality is, though, that from the receiving end, it seemed like an amazing coincidence had actually occurred. The truth is, there are thousands of people who predict the stock market, and some of them are correct sometimes, and you would expect that, because there are so many people that are making these decisions. But we also expect good results from a monkey throwing a dart, or from just flipping a coin, because we would expect that coincidence, of having information coincide just on top of each other.

The point here is that when someone advertises, says, "Well, invest with us, because, in fact, we predicted the great stock market crash of 1989," you should be wondering if that was really due to some important insight, or just lots of people guessing, or, to bring back the previous lecture, lots of monkeys typing away, and some monkey is actually going to guess that someone is going to have a crash, in particular, the market, in 1989.

Anyway, it provides a great template for us to realize that sometimes, when we experience something that seems like a coincidence, it may in fact not be.

I thought I would also share with you one of the most famous coincidences that people think about, which is the birthday question. The question is: Suppose you had 45 people at a party or an event. What is the probability that in that small group of 45 people, two of them will actually share the incredible coincidence of having their birthdays on the same day, the same month, and the same date. Well, here's a way of thinking about it. Let's just sort of think through this, and see what you think. You could argue, "Well, if there were 366 people in the room," and by the way, let's pretend that there are no leap years, so that the total number of days in the year will be 365. I apologize to those people who may have been born at the very end of February. Let's just pretend for simplicity, though, that there are only 365 days in the year.

Therefore, if we had 366 people in the room, then, from Mike's early lecture, Lecture Two, when he talked about the Pigeonhole principle, we know that they all can't have different birthdays, because there are only 365 possibilities. At least two of them must, in fact, have a birthday that matches. With 366, then, we are guaranteed that at least two people will have the same match.

Well, here's an argument. One argument would say, "Well, great. If you took half of those people, then, the chances would be cut in half." If I have roughly 180 people, then, that's roughly half of 366; then, the probability is perhaps 50-50, or half. Cut that in half again, and I have 90. The probability would be a quarter. Cut that in half again, I get 45, and the probability would be an eighth, so not very likely at all.

That might be an intuitive way of thinking about the likelihood of that event. That's certainly an intuitive way of thinking about it, but not an accurate way, because the reality is that if you have 45 people in a room, the probability that at least two of them share the same birthday, month and date, turns on to be 95 percent, so close to being certain even though we have so few people. It's so surprising, but what does that mean? It means that we need to retrain our intuition. How do we do that? Try this experiment when you are at a cocktail party or dinner party with 45 people. Go around the room and ask; or, to see it numerically and quantitatively, we have to think mathematically, so let's think mathematically for a second, and make this surprising discovery actually intuitive and commonplace.

Well, to understand this surprising probability, what we do is we look at the opposite question. Instead of saying, "What is the probability of having two people's birthdays match?"—let's look at the opposite question: What is the probability that no birthdays match among these 45 people? Let's then show that that probability is very small, which means that its opposite must be very large. Remember, in Mike's Lecture 20, he showed us that if the probability of some event is some number, then the probability of the opposite event is 1 minus that number.

Well, let's think about this now together, and reason together. Imagine lining up these 45 people in a long line, and we're going to go down it and ask them their birthdays. We want, now, to consider the probability that there are no matches at all. The first person, then, of course, has nothing to compare herself to, because she's the first person we're asking, "What day is your birthday?" That person can just say whatever her birthday is. Let's say she says, "June 16."

Fine. Now we turn to the second person, and we want to make sure that there's absolutely no match, so what is the probability that this person does not match the first person? Well, the first person has one birthday, and this person has to avoid that one date, so how many successful possible ways, how many successful possible dates are there for this person to have as his birthday, in order to avoid the birthday of the first person? The answer is 364, because that person can have any birthday at all except for June 16. What's the probability that we select one of those dates, then, arbitrarily out of the total of 365? Well, you can see here on the graphic that it is going to be 364, the number of days that don't match the first person, divided by the total number of days in the year, which is 365. That works out to be 0.9972, or 99.7 percent.

No huge surprise, right? Certainly, if you have two people, you would be shocked if they shared the same birthday, and we're seeing, that in fact, the probability is very, very high that they don't share the same birthday.

Now, let's look at the third person. Here's the first person, here's the second person, and now we ask the third person what her birthday is. Now, for her birthday to be different from the previous two people, how many successful possibilities are there? Well, two less than the total number of days in the year, so we see here that she has 363 opportunities to have her birthday on a day that differs from those of the first two people. We have to divide that by all the possible dates of the year, which are 365, and we get a fraction of 0.9945, which is roughly 99.5 percent.

Notice that it is a tiny bit less likely than the 99.7 percent, and for good reason, because now we're actually asking this person, we're demanding more. She not only can't match with this guy, but she also can't match with the first woman. In fact, we're asking a bit more, so it is becoming a bit less likely.

We go down the list. Now we ask the fourth person. Well, what would be a birthday success, meaning no match? Well, that person has to avoid this date, this date, and this date, which leaves that person with 362 possibilities out of 365, you can continue this, until you get down to the 44th person, in which case we have 322 possibilities out of the 365, and the probability there is 0.8821, or 88 percent, the probability that that person will miss all the previous people, so notice that the numbers are dropping, because we're demanding a lot more, so it's getting a little bit less likely, but still very likely—88 percent is very likely.

Finally, the last person, the 45th person, has to miss all of the previous 44 dates, so the number of birthdays that that person can possibly have turns out to be 321. Divide that by the total number of days in the year, 365, and we see 0.8794, or 87.9 percent. So that is still very, very likely, but notice that it is less likely than just two people missing each others' birthdays.

Anyway, now we have all these numbers, and what do we do with them? Well, we want them all to happen simultaneously. We want the second person to avoid the first person, and we want the next person to avoid the previous two, and we want the next person to avoid the previous three. Now, Mike actually talked about this type of phenomenon, and this type of probability in Lecture Twenty. There we saw, just by rolling dice, that the way to get all the possible combinations was to multiply these numbers together.

In fact, then, to find the probability that there is no match all among these 45 people, we have to multiply all those numbers together, in particular, 0.997 multiplied by 0.994, all the way down to 0.882 times 0.879.

Well, if you actually wanted to try this for real, which I really urge you to do. Even numerically, it is sort of hard to type in all those numbers. It's a long list of over 40 numbers that you have to type in. It is too long, so let me just tell you a little shortcut to do it. If you take the average of those numbers, you get something around 0.935, so a rough estimate of the size of that, if you don't actually want to do it out by hand, is to take 0.935, and raise it to the power 44, since those are the 44 people that have to all avoid the first person and the previous people, and that, you can actually plug into a calculator, and you will see that the answer is approximately 0.05. The product of all those numbers, then—even though individually they weren't too small, the product of those numbers turned out to be extremely small.

There's a 5 percent chance, then, that there's no match at all, so what is the probability of the opposite question, that there exists at least one match? Well, it would be 95 percent, the opposite, and that actually proves that if you have 45 people in a room, the probability that two of them share a birthday, in fact, is very high, 95 percent. It turns out, by the way, just as a little trivia fact—you can work it out the exact same way—that if you had only 23 people in a room, just 23, the probability is 50-50 that there will be a match, so that is the break-even point, in a way. It seems so counterintuitive, because it is such a small group, but the next time you are in a group of 30 people, try it, and you'll be amazed at how often there is a match. That will also help us retune our intuition to the whole situation.

These birthday matches, then, I really think, show us, and beautifully depict the disjoint between our intuition and the reality, and by doing these experiments, we do, in fact, come to grips with the reality more accurately.

You know, coincidences are more common than we usually believe they should be just by chance alone, and as we were seeing in this lecture, we need to retrain that intuition to be more accurate about what randomness does on its own. We see coincidences all around us, and what we see then, we just don't know where or when they're going to strike, but we know they're going to strike. So in some sense, the moral here is that we should really expect the unexpected every once in a while, and we saw that through a very simple process of experimentation, thinking about things quantitatively, and making it our own. The idea that something is surprising is really just a departure from our understanding of the actual reality, and is an important insight into the thinking of how we actually operate.

Well, these past two lectures really point out that unlikely things will occur all around us. Now, once we realize this truth, we see our lives in a whole new way. Our lives are really just long runs of incredibly unlikely events, all strung together, and it is sort of a wonderful way of thinking about how we journey through life, as sort of bumping around and hitting things in a sort of random fashion that makes up the stream we call our life's journey. You know, viewed in this manner, we see that each of our lives really resembles a monkey's typed version of *Hamlet*, but just with fewer footnotes.

The point is that we now see through a careful, mathematical way of thinking, through a very careful and rigorous focus on the issue at hand, that we can make sense out of things even as nonsensical as the everyday coincidences we experience, and that we should, in fact, not only expect coincidences, but we should almost embraced them. When those things happen, we should be shocked by the fact that they're happening, for example, those two people of the 45 that have the same birthday. With respect to them, an amazing coincidence has occurred. They each found someone who was born on the very same day as the other person, but even more surprising is that no coincidence at all would occur.

From the vantage point of the person who experienced the coincidence, something magical has happened, and I think that in fact, it is just a wonderful metaphor for our entire journey. Something magical really has happened. By simply just looking at things in a focused way, by really experiencing the joy of thinking, we were able to embrace realms, and worlds, and issues that seemed so foreign, both within the context of mathematics, and within the context of our lives, and our hope is that somehow, all these ideas, whose power we've seen, through these different vistas of classical mathematics, will come together and culminate in your lives.

In the next lecture coming up, our final lecture, Mike and I will come together and take a look at these thinking strategies, these life lessons, that really pull together to offer a new way of looking at everything, not just mathematics, but our own lives.

Lecture Twenty-Four
Life Lessons Learned from Mathematical Thinking
Edward B. Burger, Ph.D. and Michael Starbird, Ph.D.

Scope:

This course presented a wide range of mathematics and illustrated the different themes that mathematics embraces. The mathematical perspective can help us appreciate many aspects of our world more deeply. Mathematics is both ancient and modern. Ideas that were discovered long ago are as meaningful and intriguing today as ever. This series of lectures touched on a broad range of mathematical themes and took us on a tour of some of the more classical realms of mathematics. But mathematics is a living discipline. Every day, new mathematical discoveries are made. Whether looking at classical ideas or modern developments constructed on classical foundations, mathematics truly is a subject with no end to its concepts or its intrigue.

Outline

I. This lecture closes the course by bringing to the foreground the strategies for thinking, or "life lessons," that we have employed. Our top 10 list of these life lessons is as follows:

A. Just do it. If you're faced with a problem and you don't know how to solve it, begin by taking some action. We saw this strategy at work in the early lectures on counting and the Pigeonhole principle, as well as the later lectures on coincidences.

B. Make mistakes and fail but never give up. Mathematicians are supremely gifted at making mistakes, as is anyone involved in a creative endeavor. The key is to use the insight from your mistakes to identify the features of a correct solution to your problem. We explored this strategy in discussing the Fibonacci numbers and Sierpinski carpet.

C. Keep an open mind. If we are never willing to consider new ideas, then we can never hope to increase our understanding of the world around us. In every lecture of this course, we looked at ideas that lie beyond our everyday experience; we saw this principle at work, particularly, in the lectures on the fourth dimension and probability.

D. Explore the consequences of new ideas. This strategy pushes us to see where an idea leads and, in this way, to discover new ideas and insights. As we saw in our discussions of Fibonacci numbers, the Möbius band, and the Klein bottle, the most important work you might do on a problem comes after you have solved it.

E. Seek the essential. One of the biggest obstacles in solving real-world problems is the noise and clutter of irrelevant issues that surround them. The lectures on the collage method of creating fractals and the Golden Rectangle taught us to examine complicated situations for the essential ingredients that lie at the crux of the situations.

F. Understand the issue. Identifying and clarifying the problem to be solved in a situation is often a significant step in reaching a solution. Our discussions of random walks offered an outstanding example of this principle, as did our proof that cutting a cone with a plane forms an ellipse.

G. Understand simple things deeply. This technique may be the most potent strategy for effective thinking of all that we have employed. We can never understand unknown situations without an intense focus on those aspects of the unknown that are familiar. The familiar, in other words, serves as the best guide to the unfamiliar. We used this strategy in exploring abstract concepts and creations, such as the fourth dimension and the Klein bottle.

H. Break a difficult problem into easier ones. This strategy is fundamental to mathematics and, indeed, applicable in everyday life. Can we solve a difficult problem by, instead, looking at two simpler problems? We broke down difficult questions in our explorations of the fractal Dragon Curve and the Euler characteristic.

I. Examine issues from several points of view. We can, for example, gain new insights by looking at the construction of an object, rather than the object itself. Again, we saw this principle applied in our construction of a Möbius band, as well as our examination of *Let's Make a Deal*® and Newcomb's paradox in the lectures on probability and expected value.

J. Look for patterns. Similarities among situations and objects that are different on the surface should be viewed as flashing lights urging us to look for explanations. Patterns help us to structure our understanding of the world, and similarities are what we use to bring order and meaning to chaos. We saw this technique used in the lectures on the Platonic solids and paper-folding.

II. We hope that this course has helped you learn to view mathematics as an intriguing collection of ideas that has come to life through powerful methods of thinking. We also hope that the thought processes we have explored in this course will bring added richness to your everyday life.

Suggested Reading:

Edward B. Burger and Michael Starbird, *The Heart of Mathematics: An invitation to effective thinking*, Key College Publishing, Section 5.2, "The Band That Wouldn't Stop Playing: Experimenting with the Möbius Band and Klein Bottle."

Questions to Consider:

1. Take a completely non-mathematical issue that you would like to understand better. Apply some of the methods of thought in this course and see whether you get a new idea or insight about the issue.

2. To what extent do you feel that more effective and imaginative people are more effective and imaginative owing largely to habitually exploring their world using methods of thinking similar to those presented in this lecture?

Lecture Twenty-Four—Transcript
Life Lessons Learned from Mathematical Thinking

Ed: We hope you enjoyed these wonderful vistas of classical mathematics and now view them as we do, as some of the greatest ideas of humanity.

Mike: In this lecture, we want to bring the methods of thinking and the techniques of analysis that we've employed into the foreground. We could even view these strategies of thought as "life lessons."

Ed: Well, everyone nowadays has their favorite top 10 list, and so we thought we would end with our top 10 list of favorite life lessons that we've employed in this course. Number One is: Just do it. Sometimes when we are faced with a problem, we simply don't know what to do, we freeze like deer in headlights, because from where we are, we cannot see a complete solution. The "just do it" phrase means, to us, that sometimes we just need to take a step, even if we don't know where that step will lead. By merely taking an action, we gain insights that often become our guide to an imaginative solution. Now, where have you seen this?

Mike: Well, I think that the excellent example occurred in the lectures about counting. One thing you do with counting is just to do it, to actually take the step of counting how many or how long something would take. For example, we had this example of dropping a $100 bill, and wondering whether Bill Gates should pick it up, or folding paper 50 times, trying to see how the thick that would be, and the way that we figured these things out was to actually take the trouble to do the process, to add the numbers together. It's very simple arithmetic to do, but then we saw something special about it. The Pigeonhole principle was another example, where just thinking about doing the process of, say, putting the people into different rooms depending on how hairy they were led us to an insight. Do you have an example of that?

Ed: Well, I think when we were talking about coincidences, they were a wonderful illustration of how, just by doing it, we can actually try and build insights. When we first thought about the birthday issue, and we had this birthday question about: If we had 40 people, how likely would it be for two people to have the same birthday? At first, it sounds highly unlikely, because there are so many possibilities for birthdays. One way to change that initial impression to the more accurate impression that, in fact, it's very likely, is to try it and when you have collections of 40 people, ask, and collect birthdays, and do experiments. That allows us to rehone our intuition, and to get a firmer grasp intellectually and intuitively as to the idea at hand, in this case, the coincidence thing, which is so hard to measure. So often, we can build that insight by just trying it, by actually trying it. And with 24 people, we would expect that half the time we would have a birthday match, and half the time we wouldn't, and to experience that and enjoy that for real.

Mike: Well, let's move on to Life Lesson Number Two. Number Two is: Make mistakes and fail, but never give up.

You know, in a way, mathematicians are supremely gifted at making mistakes, but the truth is that anybody who does something creative will make mistakes. No one has ever done something truly original without first making mistakes and sometimes failing. The key thing is to take advantage of the mistakes, to see what went wrong, and then actually use that insight to do better the next time. Actually making the mistake allows us to see more clearly what features a correct solution has to have. Actually, then, mistakes and failure are invaluable strategies for guiding us to the correct answers.

Let's see if we can think of some examples. What are some examples?

Ed: One example is with the Fibonacci numbers. We started with this sequence of numbers that we found through counting, and we saw that the collection of numbers was increasing, and then, the question was: How fast are the numbers getting larger? In our first guess, we just tried something, which was, "Well, maybe they're doubling each time," and so, we started by going from 5, and we moved to 8, which was not quite right, because if we doubled 5, we would move to 10, and so, we saw that just doubling was not quite the right way of seeing how fast these numbers were growing.

That mistake, though, led us to discover a new feature about the Fibonacci numbers, in particular, that formula that actually showed how doubling one Fibonacci number relates to the previous one, because, in fact, we can compare them, and see a formula. Then, once we had that formula in hand, we were able to say, "Okay, that mistake didn't work. It just failed. What is the growth rate of the Fibonacci numbers?" We started to look at the quotients. Once we looked at the quotients, what did we see? We discovered the Golden Ratio, which led to all sorts of wonderful new discoveries not only about the Fibonacci numbers, but the Golden Rectangle, and aesthetics, all from starting off by making a mistake, by thinking the Fibonacci numbers were doubling each time, which, it turns out, they weren't.

How about you? Where do you see it?

Mike: Well, one example that we encountered was in the development of the fractal that we called the Sierpinski carpet. You remember that this was the fractal that was made with a square, and then we put a little square inside it that was a different color, and then we sought the challenge of: Could we make each of the eight surrounding squares identical to the whole picture?

We first put one square in the middle, and then, we said, "Well, no, it's not quite right, because the square on the outside doesn't have a colored spot in it." We made a mistake, and we corrected that mistake. We put a colored spot in the middle of it, and then we looked at what we had, and we said, "Whoops, no, it didn't quite happen again, because now, we have a middle square, and we have these other little squares around, so that the corner one is not the same as the whole one," and so, we had made another mistake, and we just continued over, and over, and over again, so that, in this case, in fact, we kind of made infinitely many mistakes.

Ed: Number Three on our life lesson list is to keep an open mind. You know, nothing is more important when thinking of a new idea than a sincere willingness to actually consider it. If we do not consider possibilities that seem foreign or even counterintuitive to us, then, we can never hope to rehone or reconceptualize our understanding. In essentially every lecture, we explored issues that required us, indeed, required humanity before us, to be willing to consider ideas that lay beyond our initial experiences. Now, where do we see this?

Mike: Well, one example was in the probability lecture. You might remember that we had this sort of interesting reunion scenario, where a person at the scenario said, "I have two children," and then said, "My older one is a boy," or "One of them is a boy," and at first, when we thought about the differences between those two very similar situations, it seemed obvious that the chances in each case were 50-50 that she had two boys. You wouldn't even think of asking the question—what is the distinction in the probabilities?—without being open-minded enough to consider the possibility that, in fact, the world is a little different from what you actually expected. That's an example.

Ed: One thing that we saw, I think, to illustrate keeping an open mind, is the fourth dimension. Just the whole realm of thinking about that was fantastic, and genuinely required us to keep an open mind. When we think about it, once we developed the idea of what dimension was, as degrees of spatial freedom, we quickly saw that we, in fact, were living in a three-dimensional world. We can go this direction, this direction, or this direction, so that the idea of a fourth dimension seemed impossible, because we couldn't even point to it. Wherever we point, we continue to stay in our three-dimensional space.

By keeping an open mind, though, we were all of a sudden able to analyze a geometric world that we cannot, in fact, see with our physical eyes, but that we can see with the eye of the mind, and through that, we were actually able to explore and discover new insights, not only about a four-dimensional space, but even new insights about our very at-home three-dimensional space. In fact, physicists are now considering the possibility that maybe our universe is not three dimensions. Therefore, just our familiar surroundings, somehow, are sort of not enough. In fact, we need to always keep our minds open.

Mike: Okay, we will move on to Life Lesson Number Four: Explore the consequences of new ideas. These life lessons are really strategies for coming up with new ideas, and this one is one of the most productive strategies, I think, for thinking of new thoughts, because the way to think of a new thought is to see where an idea leads. I had a mentor who was a wonderful mathematician when I was first starting at The University of Texas, and he used to tell me that the most important time to work hard on a mathematical problem was after you had solved it. And what he meant by that was when you get an idea, the idea that led to the solution, very often, a lot of new insights can be found if you take the trouble to methodically consider the implications of the new idea. Do you have some examples?

Ed: Well, I think the Fibonacci numbers certainly illustrate this idea, because we just started by looking at the surface of a pineapple, and just literally counted the number of spirals going one way, the number of spirals going the other way, and we took that number, and started counting other spirals, and other natural objects that we saw, flowers, etc., and all of a sudden, a pattern emerged that led us to define the Fibonacci numbers.

So, we have the Fibonacci numbers, and that's fine. Now, let's explore the consequences of that, and that exploration took us all over the place, from the reproductive habits of rabbits to the Lucas numbers, which is a completely different sequence, but is somehow interconnected with the Fibonacci numbers, to the Golden Ratio, to the Golden Rectangle, and even to the ability to express every single natural number as a sum of the Fibonacci numbers. All of this rich structure came out of simply exploring the consequences of that new idea, the idea of counting the spirals and finding that pattern.

Mike: Well, I have an example. When we were discussing the Möbius band and the Klein bottle, we had the idea, that to make a Möbius band, you took a rectangle and glued the edges together. Well, that was an idea. It was an idea of how we could represent some object like the Möbius band by a construction diagram, and then we took that idea, of how we represent an object by a construction diagram, and followed it through, and said, "Now, can we use that same idea in a slightly new setting?" Indeed, we could, and it led to marvelous things, including the beautiful Klein bottle, that one-sided surface.

Ed: Number Five on our list: Seek the essential. One of the biggest obstacles in solving real-world problems is the noise and clutter of irrelevant issues that surround them. An excellent guide to discovery is to take a complicated situation, and ask: What is the essential ingredient that lies right at the crux of this situation? Seeking the essential forces us to focus on the real issue, rather than being distracted by debris. Now, where did you see this?

Mike: Well, one wonderful example of this came in the collage method of creating fractals. You'll remember that this was a method where we took a picture, and then we reduced it in size and arranged it on a page, and then we took the resulting collage, reduced it in size, and put it on the page, and we repeated this process, in fact, infinitely many times. We saw that we could even get these wonderful pictures, like the Barnsley fern that looked exactly like a fern, even though it was just constructed by this pattern, and the thing that we noticed about it was that it didn't matter what picture we started with, that really, that was irrelevant. The irrelevant feature was the picture that we started with, and yet, it seemed at first that that would be the crux of the matter.

Noticing, then, that it was completely irrelevant made a beautiful example of forcing us to focus our attention on what the actual, essential feature of that situation was. Do you have an example?

Ed: You know, when we took a look at the Golden Rectangle, we first came upon it in terms of the ratio of its base to height. We saw that the ratio should be the Golden Ratio, and so we explored the Golden Rectangle, and we saw visions of it in various different architectural and art forms, but the question that really came up was: Why was this rectangle so aesthetically appealing to us? It wasn't just the definition. It wasn't that definition of base to height being the Golden Ratio. It was through the study of the villa by Le Corbusier where insight was formed. Namely, once we chopped off the square region that made up the actual villa, that was the largest square that we could remove from the villa, we noticed that the remaining rectangle, even though it was on its side, the rectangle portion that made up the patio, we saw that, in fact, that rectangular portion was another perfect Golden Rectangle.

What was the insight that we discovered? The insight we discovered was that if you take a Golden Rectangle and remove its largest square, the remaining smaller rectangle has the same proportions. Namely, it is a Golden Rectangle. Then, we realized that that is the only rectangle that has that feature, so it's not just the definition. But it was through the definition that we actually sought the essential feature of that rectangle, namely, that it is the only rectangle you could possibly imagine for which, when you remove the largest square, the remaining portion is proportional to the entire sum, and that might, in fact, explain part of its appeal. But it was finding that essential feature that led to that.

Mike: Okay, let's move onto Life Lesson Number Six: Understand the issue. Very often, when we are faced with a question, we might find ourselves diligently working away, but on the wrong problem. I experience this as a professor very often. Students often come to my office, and they will say something like, "I don't know how to do this word problem." I will say, "Okay, what exactly is the question?" They will answer, "It's question number seven on page 324." They are so anxious to go about solving the problem that they haven't really spent the time to understand what the question is. In fact, very often, once you have understood the question, 90 percent of that question is solved. You are 90 percent of the way to the solution when you really have clarified what the question is. What's an example?

Ed: In talking about randomness, it's a beautiful example, because it sounds like there's no way to really understand that, and wrap your mind around it. In particular, if we think about the random walk scenario that we considered, where we start on a line, say, at the origin, and we move one step to the right or left, depending on how the flip of a coin goes—if it's heads, we go one way, and if it's tails, we go the other way, and from that new point, that new position, we flip again, and we keep traveling, and we're generating this random walk. The question was: Is that random walk transient? If we start at the origin, will we at some point return back home?

At first, it sounds like an impossible question to answer, because there are so many things to involve, and with the randomness of it, it seems impossible. The key is to understand the issue, and when we thought about it, we gained several important insights. Number one is that returning back to where you start is nothing more, in this scenario, than flipping just as many heads as we flip tails, which immediately showed us that we could never return back after an odd number of steps. We have to take an even number of steps to return home.

The other question is: If we keep flipping a fair coin, where the chances of landing heads or tails are equal, and we keep flipping it, what is the likelihood that at some point, the number of heads we flip will be equal to the number of tails? Well, it seems clear now, intuitive, that, in fact, that should happen. It's happened many times, so in fact, we should return home, and all of a sudden, this question about randomness that seems rather impossible becomes very clear. Once, in fact, we understand the issue, and realize that insight. How about you?

Mike: Well, let me give a geometrical example. When we were talking about the ellipse, we said, "If you take a cone, and you cut it with plane, you get an ellipse." Well, people know that, and they say,

"Well, yes, that's an ellipse. Okay, we're done." The question really is, though: Why is it that when you cut a cone with a plane, that that oval-shaped curve actually has this very interesting property that there are two points such that for every point on this curve, the sum of the distances to those two points is the same no matter what point you choose around the entire curve.

The issue associated with the ellipse question was: What is the question? The question is how to characterize the form that you get when you pass a plane through a curve, and see it in two different ways, one as the intersection of a plane with the cone, and the other as a set of points so that the sum of the distances to those two points is a constant. That's a good example.

Ed: Our Number Seven Life Lesson is: Understand simple things deeply. This technique may be the most potent strategy for effective thinking ever. It arises again and again, both in and more importantly, beyond mathematics. If we are trying to understand the unknown, we are always stumped, because we do not know what to do. However, one action we can take is to revisit those issues that are familiar and ordinary, and focus ourselves to understand them in a far deeper way than we normally do. The familiar, understood deeply, is the best guide to the unfamiliar and the unknown. Mike, where did we see it?

Mike: Okay, one example is the Klein bottle. The Klein bottle is something we never would even have been able to begin to think about. It is a very abstract object. It can't actually physically be built, except in the fourth dimension. It's an object of the mind, but the way that we got to it was to start with something much, much simpler, namely, one thing we started with was the torus, the boundary of a donut. That object is much simpler. And what we did was that we came to understand that, namely, as folding up a piece of paper in two ways to create this torus, so that we're able to take that understanding of the familiar torus, and then apply a variation of it. Because we understood it so well, we could do a variation that led us to the construction of this beautiful Klein bottle.

Ed: Another very powerful example is just that realm of the fourth dimension that we talked about, the idea that somehow, we can wrap our minds around a world that we cannot perceive, or even see, is remarkable, and at first, seems impossible, and what was the key? The key was to understand simple things deeply. We argued by analogy. We didn't try to understand a four-dimensional world that we can't see. Instead, we retreated. We retreated back to smaller dimensions; two dimensions, the two-dimensional world of this plane, the surface of a table, or the one-dimensional world of a line, and we explored those ideas so deeply that we actually built new insights into the line and the tabletop, and then, actually, new insights into our three-dimensional space, and that allowed us to actually move far beyond, into the fourth dimension, all by understanding simple things deeply, in this case, by arguing by analogy.

Mike: Number Eight: Break a difficult problem into easier ones. This is a life lesson that for mathematicians is ingrained to the point of instinct, and maybe mathematics displays the strategy most clearly, because numbers let us quantify how difficult some problems are. Nevertheless, this strategy is extremely fundamental. If we have a complicated problem, or a difficult problem, it is absolutely basic to ask the question: Can we instead do two simpler problems, and then combine their answers to solve the original problem? It's easier to solve two easy problems than it is to solve one hard one. Do you have an example?

Ed: Well, I think one great example is in our investigation of the Dragon Curve, that fractal that looks like a dragon, which actually covered and tiled the entire plane. When we first looked at it, we appreciated its infinite complexity, but we also appreciated the fact that it seemed impossible to understand exactly how you would actually build or describe that thing. What we did was that we didn't answer that question. We broke it down, and we looked at easier ones, and what we saw was that through the very simple act of folding a sheet of paper repeatedly, by unfolding that act, and making sure that all those creases were exactly at 90 degrees, in the limiting process, we produced that amazingly complicated Dragon Curve. How about you?

Mike: Well, one that comes to my mind is the discussion of the Euler characteristic. You may recall that the Euler characteristic occurred when you drew a doodle on a piece of paper, and you counted the vertices, the edges, and the regions, and you discovered that the vertices minus the edges plus the regions always are equal to 2. The way that we got there, though, was little by little. If we started with a really complicated curve, it was not at all obvious why this relationship of vertices, edges, and the areas should be true, but what was very simple was to break the problem into little tiny pieces, namely, that you just would draw one arc. Well, what drawing one arc does is that you can easily analyze what is going to happen. The first arc is easy to analyze, and then, you can analyze what happens when you just add one more edge, and then one more edge, and then one more edge, so that you have taken what you are able to do and deal with very complicated final situations, just by taking it step-by-step.

Ed: Number Nine on our list of life lessons is: Examine life issues from several points of view. We can actually force ourselves to look at issues from several points of view; we can take an issue, and examine it quantitatively, visually, or physically. We can take a representation of a question in different ways. We can gain new insights and new perspectives by looking at the construction that produced an object, rather than the object itself.

Now, all of these are methods for looking at issues from different points of view, and such exercises actually allow us to see the issue at hand from a different and distinct angle, and often, that angle can shed new light on the issue, and lead to a novel solution. Now, where have we seen that?

Mike: Well, I actually have two of them this time. One of them is the Möbius band, where looking at it, the construction process, was a representation that led us a long way. I want to look at another one, also, though, which is the *Let's Make a Deal*® game in probability. Now, this was a fun one. We were pretending we were in a game show, and we had to choose a door, and then, the host opened a door, and we had to decide whether to switch or to stick to our original guess, and a good way to be able to change our intuition was to look at the problem in a different way, namely, to take a very extreme example, where we have a billion doors, and then see if it became obvious what happened. That was a case where a change in representation of the problem, a change in how we looked at the problem, allowed us to see something quite clearly that was not so clear in its original perspective.

Ed: One thing that we also saw together was Newcomb's paradoxical situation, a scenario where we walk into a room, and there are two boxes. In one box, we see $1000, and one box is completely covered, and we don't know. It's opaque. It could contain

$1 million, or nothing, and we have to make a decision. What do we decide, then, given all the information that we have from Newcomb's paradox? Well, we saw, in fact, that it genuinely is a paradox, by examining it from several points of view.

If we think of it in terms of an expected value question (a question where we repeat the experiment again, and again, and again), what we see is that the value of taking just the $1 million box was quite high, and so, maybe that's the way we should proceed. On the other hand, we only get one shot in that door, and we get to do one or the other, and so, if we believe in free will, then, in fact, it's a done deal. No matter what that scientist did, there's either $1 million in that box or not, so perhaps, we should just take them both.

The idea of examining issues from several points of view, then, really brings focus and clarity to things, even when there is no definitive answer, because, in fact, in this case, it is a paradoxical situation, but to appreciate that paradox, we had to look at these things from different points of view.

Mike: Well, we're finally at Number Ten. Life Lesson Number Ten: Look for patterns and similarities. When we come across apparently different objects or ideas that seem to somehow display similarities, or patterns to them, then what we have really found is a flashing light that beckons us to seek an explanation. Patterns help us to structure our understanding of the world, and similarities among things are what we really use to combine the chaotic view of the entire world into meaning. By seeking patterns, and overtly seeking to find the similarities in their existence, and their potential, we turn our minds to a method of organizing our experience to make it meaningful. What's an example?

Ed: Well, one wonderful example is with the paper-folding. We started with a very simple process of taking a sheet of paper, and repeatedly folding it on top of itself, again, and again, and again. When we unfolded it, and looked at the valleys and ridges that were caused through the folds, we saw this chaotic mess that just seemed absolutely impossible for us to really fathom and understand. What did we do? Well, we just ignored that, and instead, we looked for a pattern, so we started back at the beginning, and we slowly devised a pattern, and that chaotic mess all of a sudden became tamed, and we had structure. We were even able, then, to predict what future folds would look like, even without physically doing them. This idea of searching for a pattern, then, through this chaotic mess of the paper-folding, was a very powerful way to not only understand the paper folding, but then appreciate it, find the Dragon Curve, and also understand better an example of automata theory. How about you? Where did we see it?

Mike: Platonic solids are a wonderful example of this. We first employed the idea of counting to examine these beautiful Platonic solids in more detail, and then we made this chart that showed the vertices, the edges, and the faces of every one of the five Platonic solids. When we looked at that collection of numbers, we saw some similarities. They jumped out at us. The number of vertices of the cube was the same as the number of faces of the octahedron, and vice versa, and that was the shining light that led us to this really beautiful idea of the duality of the Platonic solids. That was a wonderful example, that by looking for patterns, and following through, we discovered something really interesting.

Ed: Well, we hope that you see mathematics as a collection of wonderfully large and intriguing ideas that come to live through a powerful way of thinking.

Mike: By the way, we also hope that the thought processes that we've encountered here can in some way bring added richness to your everyday lives.

Ed: Our hope is that through the ideas we have explored here together, in some way we have touched your lives. You know, the invisible world of the mind can open our eyes to new and intriguing features of our world. One of the great features of mathematical thinking is that it is an endless frontier. The further we travel, the more we see over the emerging horizon. The more we discover, the more we understand what we have already seen. How many more ideas are there for us to explore and enjoy? Enough to fill a lifetime.

Mike: Thank you for joining us for our course.

Timeline

30,000 B.C.	Palaeolithic peoples in central Europe and France recorded numbers on bones.
3000 B.C.	The abacus was in use in the Middle East and around the Mediterranean. A different sort of abacus was in use in China.
2000 B.C.	Babylonians developed a base-60 counting system with extensive calculational capabilities.
540 B.C.	Pythagoras founded his school and proved the Pythagorean theorem.
387 B.C.	Plato founded the Academy.
300 B.C.	Euclid presented the axiomatic method in geometry in his *Elements*.
225 B.C.	Apollonius described the geometry of conic sections.
A.D. 150	Ptolemy had many important results in geometry. His theories in astronomy were accepted for the next 1000 years.
1200	Fibonacci brought knowledge of Islamic mathematics to Italy.
1336	At the University of Paris, mathematics was made a mandatory subject for a degree.
1489	The first appearance of + and − signs occurred in an arithmetic book in German by Widman.
1500	Leonardo da Vinci kept extensive notebooks of his interests in mathematics, anatomy, engineering, and art, among other topics.

1543	Copernicus published his work stating a heliocentric model of the orbits of planets.
1600	Kepler and Galileo did work on motion and planetary motion, describing those mathematically.
1629	Fermat redeveloped Apollonius's work on conic sections.
1635	Descartes discovered Euler's theorem for polyhedra, $V-E+F = 2$.
1665	Newton invented the derivative and the integral.
1676	Leibniz invented the differentials of basic functions (independently of Newton).
1736	Leonhard Euler began the field of topology when he published his solution of the Königsberg Bridge problem.
1758	Fifteen years after his death, the appearance of Halley's comet on December 25th confirmed Halley's predictions.
1801	Gauss published the book *Disquisitiones Arithmeticae*, which included his proof of the fundamental theorem of algebra.
1815	The *log-log* slide rule was invented by Peter Roget (the author of *Roget's Thesaurus*).
1817	Johann Karl Friedrich Gauss began working on non-Euclidean geometry and laid the foundations of differential geometry.

1827	Möbius's work on analytical geometry, *Der barycentrische Calkul*, became a classic and included many of his results on projective and affine geometry.
1854	George Boole developed Boolean logic in *Laws of Thought*.
1874	Cantor invented set theory.
1879	Kempe published his "proof" of the four-color theorem.
1881	Used in set theory, Venn introduced his *Venn diagrams*.
1900	Hilbert posed 23 problems at the Second International Congress of Mathematicians in Paris as a challenge for the 20th century. Many have been solved to date; they are considered milestones.
1905	Einstein published his simple, elegant special theory of relativity.
1913	Ramanujan first wrote Hardy from India.
1936	Turing developed the concept of a Turing Machine.
1970	Mandelbrot coined the term *fractal*.
1976	Appel and Haken showed that the four-color conjecture is true using 1200 hours of computer time to examine about 1500 configurations.
1995	Andrew Wiles, with help from Richard Taylor, published a rigorous proof of Fermat's last theorem.

Glossary

Algorithm: A recipe or procedure for solving a problem in a finite number of steps.

Barnsley's fern: A natural-looking, fern-like fractal that can be created by a simple collage method process, created by Michael Barnsley.

Borromean rings: A configuration of three rings with the property that, if any one of them is removed, the other two can be separated, but with all three present, they are inseparable.

Brownian Motion: A phenomenon of random movement, such as particles in fluid.

Buffon's Needle: Buffon's Needle refers to the experiment of dropping a needle randomly on a lined piece of paper, where the needle is as long as the distance between the lines. The probability that the needle lands on a line gives a way to compute π using randomness.

Common factor: Given two integers a and b, a common factor is an integer k such that k divides evenly into both a and b.

Conic sections: The conic sections are the parabola, ellipse (which includes the circle), and the hyperbola. These are known as conic sections because they are formed by the intersection of a plane with a cone.

Dimension: One of the degrees of freedom in space. For example, a line has one dimension, a plane two, and a room appears to have three dimensions.

Duality: Two objects exhibit duality if higher-dimensional features of one correspond to lower dimensional features of the other and vice versa. For example, the cube and octahedron exhibit duality, because the number of vertices of the cube equals the number of faces of the octahedron and vice versa.

Euler characteristic: Given a graph or doodle, the Euler characteristic is the number $V-E+F$, where V is the number of vertices, E is the number of edges, and F is the number of faces. For any connected drawing in the plane or on a sphere, the Euler characteristic always equals 2.

Expected value: The average net gain or loss that one would expect if a given probabilistic event occurred many times.

Fair game: A game where the expected value is zero.

Fibonacci numbers: Members of the Fibonacci sequence.

Fibonacci sequence: The sequence of numbers starting with 1, 1, where each subsequent number is the sum of the previous two. Written as a recursive formula, the nth Fibonacci number $F_n = F_{n-1} + F_{n-2}$. So the Fibonacci sequence begins: 1, 1, 2, 3, 5, 8, 13, 21, 34, …

Focus (foci): The foci of an ellipse are the two fixed points such that the ellipse is the set of all points such that the sum of the distances to the two foci is a fixed given value.

Fractal: An infinitely detailed pattern or picture with the quality of self-similarity; on any scale the shape resembles itself.

Geometry: The branch of mathematics that studies the visual world and abstractions of it such as lines, angles, points, surfaces, and solids.

Hypotenuse: In a right triangle, the side that does not meet the right angle.

Integer: A whole number (positive, zero, or negative); …-2, -1, 0, 1, 2…

Irrational numbers: A real number that is not rational. Examples: π and the square root of 2.

Iterate: To repeat.

Klein Bottle: An elegant one-sided surface formed by connecting the two ends of a cylinder in such a way that the resulting solid has no inside or outside. It can only be portrayed in three-dimensional space by introducing a self-intersection. It can be drawn completely in four-dimensional space.

Koch curve: A self-similar fractal that is infinitely jagged and has infinite length.

Logarithm: The exponent to which a base must be raised in order to produce a given number. For example, if the base is 10, then the logarithm of 1000 is 3, because $10^3 = 1000$.

Lucas sequence: The sequence formed with the same rule as the Fibonacci sequence, only starting with the first two numbers as 2 and 1. The first few terms in the Lucas sequence are 2, 1, 3, 4, 7, 11, 18, 29, 47, …

Menger sponge: A self-similar fractal in 3-dimensional space formed by a sculptural collage iterative algorithm that involves replacing a cube by 20 smaller cubes.

Möbius band: A surface having only one side, constructed by inserting a half-twist into a cylindrical band.

Natural numbers: The positive whole numbers: 1, 2, 3…

Newcomb's paradox: A paradoxical situation where an individual is given a choice between one box of money or two boxes of money. The specifics of the situation make each decision appear correct, even though they are opposite decisions.

Orientability: The quality of a surface of being able to have a sense of clockwise that is consistent over the whole surface.

Perfect number: A number that equals the sum of its factors. For example, 6 = 1+2+3.

Planar objects: Objects that can exist in the plane. For example, a square is a planar object, while a cube is not.

Plane: Sometimes called the Cartesian plane, the flat space that consists of the set of all points with two coordinates (x, y).

Platonic solids: The five solids whose faces are congruent, regular polygons and each of whose vertices has the same number of incident edges. The Platonic or regular solids are the tetrahedron, cube, octahedron, dodecahedron, and icosahedron.

Polygon: A closed curve that is made from a finite number of straight line segments.

Probability: A fraction between zero and one that describes how likely an event is. For events with equally likely outcomes, it equals the quotient of favorable outcomes to possible outcomes.

Pythagorean theorem: In a right triangle, the square of the length of the hypotenuse is equal to the sum of the squares of the lengths of the other two sides.

Pythagorean triangle: A right triangle with integer-length sides, for example, a 3, 4, 5 triangle.

Rational numbers: A number that can be written as a quotient of integers, with a nonzero integer in the denominator.

Real numbers: All the decimal numbers, which together comprise the real line.

Regular solids: Synonymous with Platonic solids.

Sierpinski carpet: The fractal resulting from an iterative, collage process that starts with a square and repeatedly replaces the collage at each stage with eight reduced copies surrounding a central square to produce the collage of collages at the next stage.

Sierpinski triangle: A self-similar two-dimensional shape formed by a fractal algorithm involving replacing a triangle with three triangles.

St. Petersburg paradox: Suppose a fair coin is tossed, and you receive $2 if it lands heads up for the first time on the first flip, $4 if it lands heads up for the first time on the second flip, $8 if it lands heads up for the first time on the third flip, and so on. The expected value of this game is infinite, yet how much would you pay to play?

Symmetry: A symmetry of a figure is a rigid motion of the figure or shape that results in the figure looking the same as before the motion. For example, rotating an equilateral triangle by 120 degrees is a symmetry of the triangle.

Tangent: A line is tangent to a circle or a sphere if it just grazes it, touching at one point only.

Topology: The branch of mathematics that studies properties of shapes and solids that are unchanged by elastic motions, like twisting or stretching. For example, an inner tube and the surface of a doughnut are topologically equivalent.

Torus: A surface having the shape of the boundary of a doughnut.

Turlng Machine: An abstract machine, conceptually analogous to a computer program. Fed a tape, the machine reads the tape, one entry at a time, erases the entry, and prints an output based on a finite list of rules.

Vertex (vertices): In a geometrical object with edges and faces, a vertex is a point at the end of an edge. For example in a regular solid, the vertices are the points where the edges come together.

Zeckendorff decomposition: Writing an integer uniquely as the sum of Fibonacci numbers.

π: Greek letter denoting the value 3.1415926… equal to the ratio of the circumference of a circle to its diameter.

Biographical Notes

Cantor, Georg (1845–1918). Cantor was born in Denmark but spent most of his childhood in St. Petersburg, Russia. At age 11, his family moved to Germany, where he remained for much of his life. He completed his dissertation in number theory in 1867. Later, he developed the notions of set theory and precise notions of infinity, including the concept of different-sized infinite collections. He suffered from bouts of depression and mental illness and eventually died in a sanatorium. Combined with the fact that many of his ideas were not accepted by the mathematical community of his day, Cantor is a somewhat tragic figure in mathematics.
http://www-gap.dcs.st-and.ac.uk/~history/Mathematicians/Cantor.html
http://dbeveridge.web.wesleyan.edu/wescourses/2001f/chem160/01/Who's percent20Who/georg_cantor.htm

da Vinci, Leonardo (1452–1519). At the age of 15, Leonardo moved to Florence to become an apprentice under the artist Andrea del Verrocchio. He kept highly detailed notebooks on subjects as varied as anatomy and hydraulics, as well as artistic sketches. Much of what we know of his life today is from these notebooks, as well as tax records. In 1495, he began painting *The Last Supper*. When Fra Luca Pacioli, the famous mathematician, moved to Milan, the two became friends, and da Vinci began to develop his interest in mathematics. In 1502, he left Milan to begin work as a military engineer and became acquainted with Niccolo Macchiavelli. He returned to Milan in 1506, where he became involved in hydrodynamics, anatomy, mechanics, mathematics, and optics. It is estimated that he worked on the *Mona Lisa* around 1505.

Escher, Maurits Cornelius (1898–1972). Dutch graphic artist. Although his family attempted to steer him toward a path of architecture, Escher struggled and eventually felt that a graphic arts program was a better fit. Encouraged by Samuel Jesserum de Mesquita, he experimented with woodcutting designs and printing techniques. Several times throughout his career, Escher involved himself with mathematicians. Unable to quite grasp the abstractions, he processed their work on an artistic level. Polya's paper on plane symmetry groups was quite influential. Escher later became friends with Coxeter and incorporated Coxeter's work on hyperbolic

tessellations into several pieces. He achieved world fame over the course of his lifetime.
http://www-gap.dcs.st-and.ac.uk/~history/Mathematicians/Escher.html
http://www.cs.ualberta.ca/~tegos/escher/biography/

Euclid (c. 325 B.C.–265 B.C.). Mathematician of Alexandria, Egypt. Very little is known of Euclid's life, other than his major achievement, *Elements*, a compilation of 13 books. In it, much basic geometry and number theory is built up using the axiomatic method; beginning with definitions and five postulates, Euclid then rigorously proves each statement as a result of previous statements. Much of it remains central to the study of mathematics today. In the 19th century, the parallel postulate was reinvestigated: that "one and only one line can be drawn through a point parallel to a given line." Dropping this condition led to the development of non-Euclidean geometries.
http://www-gap.dcs.st-and.ac.uk/~history/Mathematicians/Euclid.html
http://www.lib.virginia.edu/science/parshall/euclid.html

Euler, Leonhard (1707–1783). Swiss mathematician and scientist. Euler was the student of Jean Bernoulli. He was professor of medicine and physiology and later became a professor of mathematics at St. Petersburg. Euler is the most prolific mathematical author of all time, writing on mathematics, acoustics, engineering, mechanics, and astronomy. He introduced standardized notations, many now in modern use, and contributed unique ideas to all areas of analysis, especially in the study of infinite series. He lost nearly all his sight by 1771 and was the father of 13 children.

Fermat, Pierre de (1601–1665). French lawyer and judge in Toulouse, enormously talented amateur mathematician. He worked in number theory, geometry, analysis, and algebra and was the first developer of analytic geometry, including the discovery of equations of lines, circles, ellipses, parabolas, and hyperbolas. He wrote *Introduction to Plane and Solid Loci* and formulated the famed "Fermat's Last Theorem," as a note in the margin of his copy of Bachet's edition of Diophantus's *Arithmetica*. He developed a procedure for finding maxima and minima of functions through infinitesimal analysis, essentially by the limit definition of derivative, and applied this technique to many problems including analyzing the refraction of light.

Fibonacci, Leonardo (c. 1175–1250). Italian mathematician, the "greatest European mathematician of the Middle Ages." He began as a merchant, like his father, a customs officer. This afforded the opportunity for Fibonacci to learn calculational techniques, including Hindu-Arabic numerals, not yet in use in Europe. He later traveled extensively throughout the Mediterranean coast, returning to Pisa around 1200. For the next 25 years, he worked on his mathematical compositions, of which five books have survived to modern times. He is best known for the sequence that bears his name.
http://www.mcs.surrey.ac.uk/Personal/R.Knott/Fibonacci/fibBio.html
http://www.lib.virginia.edu/science/parshall/fibonacc.html

Gauss, Carl Friedrich (1777–1855). German mathematician; commonly considered the world's greatest mathematician, hence known as the Prince of Mathematicians. He was professor of astronomy and director of the Observatory at Göttingen. He provided the first complete proof of the fundamental theorem of algebra and made substantial contributions to geometry, algebra, number theory, and applied mathematics. He established mathematical rigor as the standard of proof. His work on the differential geometry of curved surfaces formed an essential base for Einstein's general theory of relativity.

Kepler, Johannes (1571–1630). German astronomer and mathematician; mathematician and astrologer to Emperor Rudolph II (in Prague). Kepler assisted Tycho Brahe (the Danish astronomer) in compiling the best collection of astronomical observations in the pre-telescope era. He developed three laws of planetary motion and made the first attempt to justify them mathematically. They were later shown to be a consequence of the universal law of gravitation by Newton, applying the new techniques of calculus.

Mandelbrot, Benoit (1924–). Born in Warsaw, Poland, Mandelbrot's family emigrated to France when he was 11, where his uncle began introducing him to mathematical concepts. He earned his Ph.D. from the University of Paris in 1952, on a mathematical analysis of the distribution of words in the English language. In 1967, he published, "How long is the coast of Britain?" Its title illustrates the phenomenon in which an object's perimeter may have undeterminable length; the measurements vary with the scale of the

ruler, so to speak. Mandelbrot coined the term *fractal* in the mid-1970s to describe certain self-similar patterns, including the famous *Mandelbrot set*. These patterns were exploited by screen-saver programmers in the latter half of the 1990s, hypnotizing the baby-boom generation with psychedelic visuals, the likes of which had not been seen since the 1960s. Mandelbrot became professor of mathematics at Yale University in 1987.
http://www.fractovia.org/people/mandelbrot.html

Möbius, August (1790–1868). German mathematician and astronomer. He studied astronomy under several heavily mathematically inclined teachers, including Karl Mollweide, Gauss, and Johann Pfaff. Möbius contributed in analytical geometry and developed ideas that would later contribute to projective geometry and topology. He was a professor of astronomy at the University of Leipzig and observer at the Observatory, also at Leipzig. The Möbius strip is a one-sided object, which Möbius himself did not invent; however, he did describe other polyhedra with this property, and he introduced the related concept of orientability.
http://scienceworld.wolfram.com/biography/Moebius.html
http://www-gap.dcs.st-and.ac.uk/~history/Mathematicians/Mobius.html

Pascal, Blaise (1623–1662). French mathematician, scientist, and philosopher. Pascal started young and published his *Essay on Conic Sections* in 1640. He invented a digital calculator and was perhaps the second person to produce a mechanical device to help with arithmetic. He wrote scientific works concerning pressure and the vacuum. He is famous mathematically for contributing his name to Pascal's Triangle, which he wrote about and which is involved in the Binomial Theorem. He wrote a philosophical work *Pensées*, making him a rare individual who made significant contributions both to mathematics and philosophy.

Ramanujan, Srinivasa Aiyangar (1887–1920). Born in a small village southwest of Madras, Ramanujan was one of India's greatest mathematical geniuses. He was largely self-taught and gained a measure of recognition in the Madras area for his work in analytic number theory. He connected with the Western world in 1913 when he wrote Hardy, in England, and included a long list of mathematical results. His relative mathematical isolation produced the odd situation in which the compiled list contained many easy, well-known results side by side with startlingly important results. Hardy recognized his genius and arranged for Ramanujan to come to England in 1914. Ramanujan's next five years were the most productive of his life, although his health began to falter. He died of tuberculosis, within a year of returning to India.

http://home.att.net/~s-prasad/math.htm
http://www-gap.dcs.st-and.ac.uk/~history/Mathematicians/Ramanujan.html

Turing, Alan (1912–1954). Born in Paddington, London, Turing was educated at public schools, where he frustrated his teachers by his determination to follow his own ideas. He did excel in mathematics and read Einstein's papers on relativity on his own. Turing entered King's College, in Cambridge, to study mathematics. There, he developed his interest in mathematical logic. In 1936, he published *On Computable Numbers, with an application to the Entscheidungsproblem*. In this paper, he developed his best-known concept, now known as the Turing Machine. This is an abstract machine that is fed a tape. It reads the tape, one entry at a time, erases the entry, and prints an output based on a finite list of rules. It is more analogous to a computer program than to an actual computer.

Bibliography

Essential:

Burger, Edward B., and Michael Starbird. *The Heart of Mathematics: An invitation to effective thinking.* Emeryville, CA: Key College Publishing, ©2000. This award-winning book presents deep and fascinating mathematical ideas in a lively, accessible, and readable way. The review in the June-July 2001 issue of the *American Mathematical Monthly* says, "This is very possibly the best 'mathematics for the non-mathematician' book that I have seen—and that includes popular (non-textbook) books that one would find in a general bookstore." Much of the content of this Teaching Company course is treated in this book. http://www.heartofmath.com

Supplementary:

Abbott, Edwin. *Flatland: A Romance of Many Dimensions.* London: Dover Publications, 1884. For more than 100 years, this story of A. Square in two-dimensional Flatland has delighted readers of all ages. It is a story that takes place in a flat, two-dimensional world. It demonstrates mathematical concepts of dimensionality in the context of a parody of 19th-century English society and class structure.

Barnsley, Michael. *Fractals Everywhere,* Academic Press, San Diego, 2000. Michael Barnsley is one of the prominent contributors to the modern study of fractals. This textbook is a rather technical description of the mathematics behind the production of fractals. It includes many pictures of fractals and the equations that produce them as well as proofs of the mathematical theorems that underlie the study of the iterative function systems that produce life-like fractals.

Bell, E.T. *Men of Mathematics.* New York: Simon & Schuster, 1937. This book is a classic of mathematics history, filled with quotes and stories (often apocryphal) of famous mathematicians.

Blatner, David. *The Joy of π.* New York: Walker Publishing Company, 1997. This fun paperback is filled with gems and details about the history of the irrational number π.

Boyer, Carl B. *A History of Mathematics.* Princeton: Princeton University Press, 1968. This book is an extensive survey of the history of mathematics from earliest recorded history through the 19th century.

Cajori, Florian. *A History of Mathematics*, 5th ed. New York: Chelsea Publishing Company, 1991; 1st ed., 1893. This work provides a survey of the development of mathematics and the lives of mathematicians from ancient times through the end of World War I.

Calinger, Ronald. *A Contextual History of Mathematics*. Upper Saddle River, NJ: Prentice-Hall, 1999. This modern, readable text offers a survey of mathematics from the origin of number through the development of calculus and classical probability.

Davis, Donald M. *The Nature and Power of Mathematics*. Princeton: Princeton University Press, 1993. This wide-ranging book describes an array of ideas from all areas of mathematics, including brief biographies of Gauss and Kepler.

Dunham, William. *Journey through Genius: The Great Theorems of Mathematics*. New York: John Wiley & Sons, 1990. Each of this book's 12 chapters covers a great idea or theorem and includes a brief history of the mathematicians who worked on that idea.

Gies, Frances, and Joseph Gies. *Leonard of Pisa and the New Mathematics of the Middle Ages*. Gainesville, GA: New Classics Library, 1969. This book presents a look at Leonard Fibonacci and the town of Pisa in the 13th century. As mathematical knowledge from the Muslim world was brought in, the Western world saw the first new mathematical concepts since Euclid.

Hardy, G.H. *A Mathematician's Apology*. Cambridge University Press, 1942. The brilliant number theorist G.H. Hardy wrote this well-known classic. It is a memoir of his career, including the period in which Ramanujan touched his life. *A Mathematician's Apology* presents Hardy's vision of what "pure" mathematics is and how a person completely consumed by the quest for abstract mathematical insight feels about life and work.

Kanigel, Robert. *The Man Who Knew Infinity: A Life of the Genius Ramanujan*. Washington Square Press, 1991. This book is a delightful biography of one of the most brilliant mathematicians of modern times, Ramanujan. This well-documented but most readable account follows Ramanujan from the slums of India, through his time working with Hardy at Cambridge, to his early and tragic death at age 32.

Kline, Morris. *Mathematics: A Cultural Approach*. Reading, MA: Addison-Wesley Publishing Company, 1962. This survey of mathematics presents its topics in both historical and cultural settings, relating the ideas to the contexts in which they developed.

Livio, Mario. *The Golden Ratio: The Story of Phi, the World's Most Astonishing Number*. New York: Broadway Books, 2002. This book is written for the layperson and delves into the number Phi, also known as the Golden Ratio.

Scieszka, Jon, and Lane Smith (illustrator). *Math Curse*. New York: Viking Children's Books, 1995. This book is a fantastical children's story of a girl who wakes up one morning to perceive the world through a mathematical filter.

Stillwell, John, *Mathematics and Its History,* Springer, New York, 2002. This delightful book presents undergraduate level mathematics through its history as a means to unify disparate mathematical areas. Each chapter contains a biographical sketch of a mathematician and presents intriguing questions and mathematical insights in an historical context.

Tahan, Malba (translated by Leslie Clark and Alastair Reid). *The Man Who Counted: A Collection of Mathematical Adventures*. New York: W.W. Norton & Company, 1993. Originally published in Brazil in 1949, this book is a combination of clever mathematical puzzles woven through the adventures of a 13th-century Arabian mathematician.